SURVIVAL
IN THE ANIMAL WORLD

SURVIVAL
IN THE ANIMAL WORLD

FRANZ GEISER
HANS DOSSENBACH

Translated from the German by Dr Gwynne Vevers

ORBIS · LONDON

Original title 'Die Kunst zu Ueberleben'
Geheimnisse der Tierwelt
A BDV Basilius Verlag AG Co-Publication
English edition © Orbis Publishing Limited, London 1985

Printed in Switzerland
ISBN 0-85613-886-X

CONTENTS

Right: the satin bower-bird (Ptilonopus violaceus) *from Australia. The male attracts the female with an artistic and colourful courtship bower.*

Below: South American poison-arrow frogs (Dendrobates reticulatus) *secrete a highly virulent poison from numerous skin glands. Certain Indian tribes use it as an arrow poison.*

INTRODUCTION

The process of evolution implies development. For all animals and plants, the same unvarying evolutionary law is applicable. This was first comprehensively formulated by Charles Darwin in his book, *On the Origin of Species by Means of Natural Selection*, published in 1859. The raw material for evolution is provided by the enormous fruitfulness of living organisms. If each seed, each egg and each newborn organism were to survive, develop normally and produce offspring, the earth would, within a few years, become a raging sea of life, in which conditions would be intolerable. However, this does not happen, because on our planet the basic resources necessary for life – water, air, food and space – are restricted

our earth with the greatest efficiency and of withstanding the harshness of the environment – extreme cold, intense heat, flood, drought and the like – will have a chance of reaching sexual maturity. And only those organisms which become sexually mature will be able to reproduce and give their offspring the opportunity in their turn to become viable. The animals and plants of today are the descendants of millions of generations of organisms that have proved successful in the unrelenting struggle against other, less successful organisms. This millionfold selection can be seen in their physical, physiological and mental attributes. There is scarcely a tiny

Left to right: Red-kneed tarantula (Brachypelma smithi), *a Central American spider with a warning pattern. Stinging hairs with numerous hooklets penetrate deeply into human skin and cause painful burning.*

Hatching giant tortoise (Testudo elephantopus). *This hatchling has everything in the egg to develop into a viable young reptile. Eggs are a survival of the brood, and birds were not the first to reproduce in this way.*

The huge bull elephant seal (Mirounga) *has only one aim, namely to establish a harem and breed, after staking out his territory on a small island.*

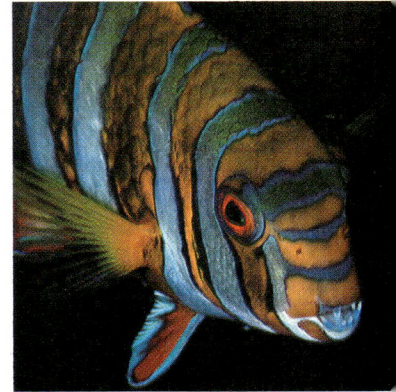

and do not increase. So each second, millions of surplus organisms die. Only a few survive. The selection, moreover, is not fortuitous. Only those with the capability of utilizing those limited resources of

hair or a fold of skin which is not of significance, which does not in some way contribute to the ability of its owner to survive in a challenging and often hostile environment. All organisms living today

Coralfishes are unrivalled for brilliance of hue and variety of pattern. The colour changes express their momentary moods.

have, in the course of thousands and millions of years, attained a degree of functional success and achievement that modern technologists can scarcely imagine, let alone emulate.

Amid this wondrous perfection, which truly arouses our admiration and respect, it may appear astonishing that we should also appreciate that animals and plants can be beautiful. Yet this very awareness of beauty, unique, so far as we know, to our own species, is itself a consequence of evolution.

This book describes some of the most interesting examples from this store of beauty and achievement. Of all the innumerable, amazing processes that have been evolved for survival, we shall discuss and illustrate those that are most truly astonishing and wonderful. There is only space here, of course, to deal with a few aspects of this subject that are no longer a mystery. We have to remember, after all, that many of the strategies which enable animals and plants to survive are probably still unknown to us and await discovery.

THE STRUGGLE FOR LIFE

The subtitle of Charles Darwin's epoch-making work on evolution is *The Preservation of Favoured Races in the Struggle for Life* – a struggle which individuals are bound to experience if they are to maintain themselves and assure the future of their kind. The phrase 'struggle for life' has given rise to many misunderstandings. It might imply, for instance, that the life of an animal consists mainly of fighting with enemies, whether of its own kind or of another species. But this is not the case. Indeed, no animal species has ever been wiped out as a result of direct fighting. Animals do not kill for the sake of killing. That, sadly, is a human phenomenon. Even predators do not exterminate their prey. The struggle, in fact, is less dramatic and more subtle. There are, for example, animals which can utilize a source of food more rapidly or more rationally than others. This means that they will enjoy an advantage over their competitors which will enable them to produce more offspring. If these offspring are similarly successful, they will increase generation by generation and thus deny their less efficient competitors a food source. As time goes by, these competitors will be pushed back into fewer and less hospitable

Below: eye of a palm gecko (family Gekkonidae) from New Guinea. Geckos, which are mainly active at night, can close their vertical pupil in bright daylight to a narrow slit. On the other hand, when the animal hunts insects at night, the pupil opens and almost fills the whole eye.

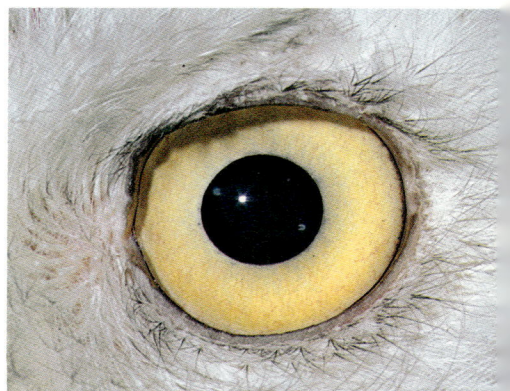

Right: eye of a snowy owl (Nyctea scandiaca). Sharp and extremely well-adapted eyes enable the snowy owl to hunt in the long arctic twilight and at night in summer.

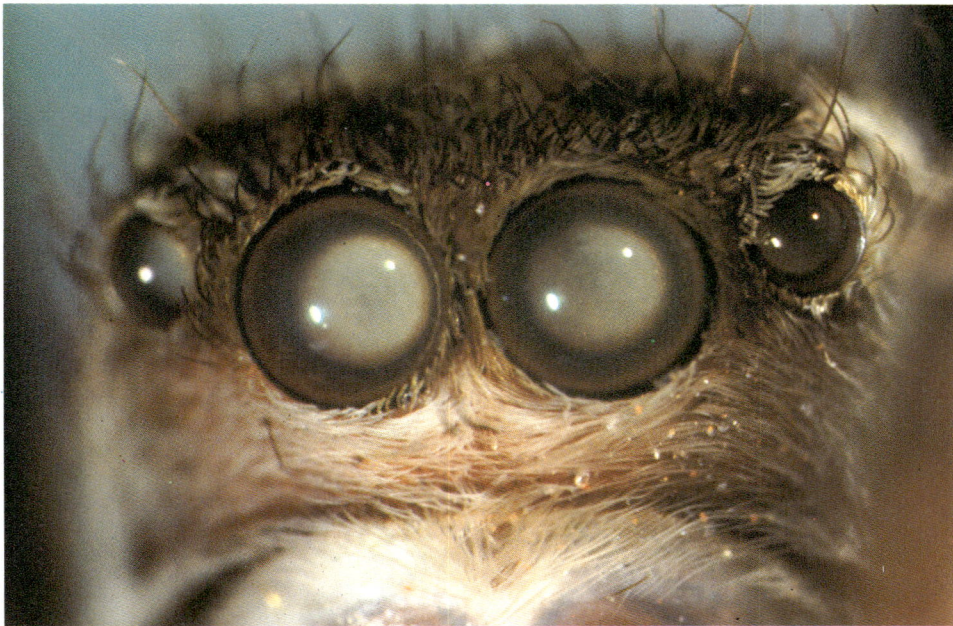

areas, and may become extinct if they are unable to adapt properly to their new surroundings. The degree of specialization also appears to play an important role in the extinction of certain animal species. Extreme specialists are dependent upon very closely defined environmental conditions. The giant anteater, for example, with

catastrophe for the species. On the other hand, a less specialized animal could switch to another food source, and so survive.

RUSES AND STRATAGEMS

How, therefore, do animals, threatened by natural forces, op-

its powerful hooked claws and long sticky tongue, is specialized for feeding only on ants and termites. If the latter suddenly vanished from the surface of the earth, the anteater would be denied its basic food and this would spell

Above: the dragonfly (Crocothemis erythrea). The compound eye of this dragonfly consists of several thousand facettes. It is essential that the eyes of this masterly aeronaut should be well equipped to detect rapid movements.

9

Right: dragonfly (Libellula fulva). Unlike other insects, dragonflies can move their front and hind wings independently of one another. This makes these rapid fliers particularly manoeuvrable.

Below left: green tree-viper (Trimeresurus), an Asiatic relative of the American rattle-snakes. Like the latter, it can use its pit organ to detect the heat given out by prey.

Below centre: South American poison-frog (Dendrobates tinctorius). The strikingly coloured pattern of this frog warns against poison. Skin poisons protect the frogs from predators and also from bacteria and fungi in the hot, humid rain forests.

Below right: bats such as the common noctule (Nyctalus noctula) first appeared about 50 million years ago. This species, like many others, is a remarkable flier, covering several thousands of miles a year.

pressed by competitors and hunted by enemies, actually manage to survive and to breed? There is, of course, no single and simple answer to this question. Each animal species finds itself in a particular situation and has to respond in its particular way. Obviously, the body of an insect has different capacities and different limitations from the body of an elephant. A polar bear basking on an ice floe in the Arctic Ocean does not have to cope with the same problems as a Malay bear in the tropical forests of South-east Asia.

an ingenious miscellany of ruses and stratagems.

A COMMON ORIGIN

It is generally believed that all life on this planet can be traced back

Far right: Dalmatian pelican (Pelecanus crispus). Birds appeared more than 100 million years ago. Thanks to the revolutionary development of feathers, they have become masters of the air.

This wealth of varied requirements and capacities has compelled each species to pursue its own path. The result of this diversity of ways and means is a fascinating array of tools and weapons and

to a common origin. In spite of obvious differences, the microscopically tiny bacteria and the massive 30-metre blue whale have probably evolved from a single, inconceivably remote ancestor. The most persuasive argument in support of this theory is supplied by the genetic code, which is the same in all living organisms. The genetic code is the molecular basis of heredity, a kind of elaborate building plan of the body, inscribed in the nucleus of every single cell. Whether a young animal comes into the world with

immediately after fertilization, when the genetic information of egg and sperm has become combined. A unique, never previously conceived individual then comes into existence. The secret message that makes this possible consists of four different molecules, the building blocks of deoxyribonucleic acid (DNA). The genetic alphabet comprises four 'letters' – A (adenine), T (thymine), C (cytosine) and G (guanine) – and their changing sequence and combination spell out the code that conditions the formation of a large number of different proteins (albumen molecules) from twenty types of amino acid. Finally, these proteins, in the form of enzymes and other substances, determine the development of the body.

Below left: Peruvian swallowtail (Iphiclides). Swallowtails are characterized by the long tails to their wings. This enables them to glide for long distances without beating their wings.

Below centre: Masai giraffe (Giraffa camelopardalis). Specializations can cause their problems. The tall body enables a giraffe to pluck leaves from trees, but it makes drinking rather difficult.

the muzzle of a dog or with the jaws of a scorpion has already been fixed prior to fertilization in the nucleus of egg and sperm. The precise form that the dog's snout will take is, however, determined

According to the information received from their molecules, these proteins determine whether the new organism becomes a bacterium, a blue whale – or a human being.

Above: lar gibbon (Hylobates lar). Type of movement is related to bodily structure. The gibbon's long arms make it a marvellous acrobat.

11

Below left: Red kangaroo (Macropus rufa) *with hind limbs that can carry it, by hopping, at speeds of 20 km per hour, using much less energy than a galloping quadruped.*

Below centre: grasshopper (order Orthoptera). *Although grasshoppers are in no way related to kangaroos, their legs likewise make prodigious leaps possible. Zoologists call this a case of convergence.*

Below right: Common vole (Microtus arvalis). *Much feared by farmers but valued as food by birds of prey, voles can reproduce at an alarming rate when conditions are right. The incisor teeth are used for feeding and digging.*

SENSE AND PERCEPTION

Movement, mating, the quest for food, the instinct to run away: life consists of a series of interactions with the living and the non-living environment. And for such interactions to take place, the basic requisite is the ability to perceive that environment. But what constitutes an animal's environment? That is not an easy question to answer. For example, a landscape that is familiar and beautiful to a human will almost certainly not be seen and appreciated as such by another animal. We take pleasure from the scent of garden flowers, whereas certain animals may, for all we know, derive a similar sensation from a heap of dung. Sensations, feelings, cannot be scientifically analysed and defined even with the aid of modern technological marvels such as electronic microscopes and electroencephalograms. Perception, on the other hand, is something different. We can be quite certain that this varies for every animal. Yet no animal can possibly perceive everything. Perception depends on the type and comparative development of the sense organs.

There are ticks, for example, that wait for months at a time on a twig in the undergrowth until a wild animal (or an inquisitive zoologist) passes directly beneath them. Then they let themselves fall, pierce the intruder's skin with

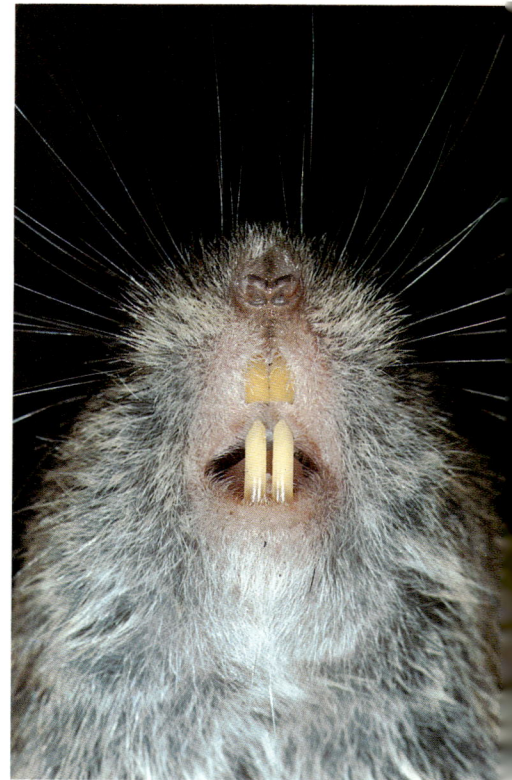

their mouthparts, and start to suck its blood. The tick has very few sense organs, but it does possess an extremely acute sense of smell, which reacts to butyric acid. The odour of this acid, which occurs in the animal's sweat, tells the tick

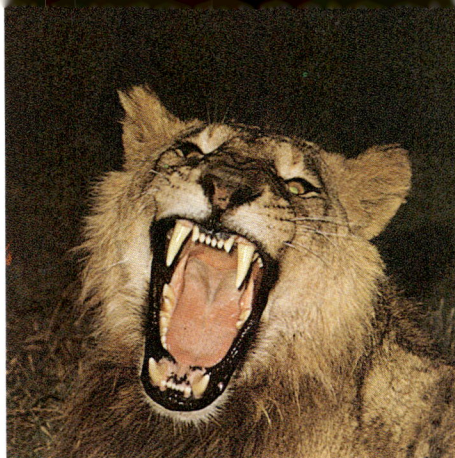

that there is something warm-blooded in the vicinity. It would be wrong to assume, therefore, that the tick perceives the world as an assembly of different animal species. Its perception would appear to depend simply on the presence or absence of butyric acid.

Human beings, nevertheless, also have their limitations. Unlike many insects, we are unable, for instance, to perceive the polarization pattern of the sky. Nor are we capable, like rattlesnakes, of distinguishing infra-red rays. Our ears cannot make out ultra sounds, which bats and dolphins habitually use for orientation. In recent times, our perception of our spatial environment has been considerably extended by certain technological advances. Even so, our awareness of the immediate surroundings is

LIFE-SAVING MECHANISMS

Perception is only one aspect of the struggle for life. Animals must also be able to influence their environment, and they have a wide range of 'tools' or mechanisms in order to achieve this. Movement, of course, is absolutely vital, so important as a rule that an animal's whole body structure may be influenced by the type of movement it adopts. Consider, for instance, snakes, kangaroos and birds, with their aerodynamic body form. Then there are animals that move in strange ways: squids and some

Left: The fangs of a predatory cat, such as a lion, are pointed and conical. They are adapted for killing prey by piercing between the neck vertebrae.

Below left: grass snakes (Natrix natrix) *alarm many people, but they are completely harmless. They feed on frogs, toads, small fishes and mice, which are swallowed whole.*

Below centre: rosy pink flamingo (Phoenicopterus ruber roseus). *The bill, furnished with fine lamellae, is held upside down to filter tiny crustaceans, algae and protozoans out of the water.*

Below: Nile crocodile (Crocodylus niloticus). *The conical teeth are well adapted for seizing prey, such as an unsuspecting antelope or gazelle, or even a giraffe, at the water's edge.*

still based primarily on the senses of vision and hearing rather than smell. By and large, therefore, our perception remains comparatively restricted, although in quite a different way from that of the blood-sucking tick.

dragonfly larvae use the principle of jet propulsion to move through the water. Certain insects are so light that they can be supported on the water surface. They alter the surface tension by releasing detergents behind them, so that the unchanged tension in front of them pulls them forwards. As an exception to the rule, there are some animals which have lost the very ability to move. Corals and sea-anemones are sessile animals, fixed to the substratum, yet in their larval stages they are mobile, thus ensuring their distribution.

The feeding apparatus of an animal, whether it be a mouth, a jaw or a beak, usually provides the zoologist with an indication of the type of food that is eaten. Thus predatory carnivores are endowed with dagger-like fangs, animals that have a fish diet possess conical teeth, those that filter food are equipped with sieve-like structures

The skin also serves at times to communicate various visual signals, such as warning colours, camouflage patterns and messages designed to indicate a readiness to mate, as a result of chemicals produced by the cutaneous glands. Nor is the skin surface necessarily bare. It may be covered by scales, spines, armour plating and so on. Other animals are furnished with organs that help them to dig; and spiders possess silk glands that enable them to spin their web.

MENTAL ATTRIBUTES

The sense organs of an animal must be able to convey information to the various organs of the body, instructing them what to do.

inside the mouth, and many plant-eaters have grinding teeth, rather like millstones.

The skin has several functions. It protects the animal against infection, desiccation, heat and cold.

Behaviour patterns vary enormously, ranging from the wholly automatic, instinctive processes characteristic of many so-called lower animals to the more complex examples of higher intelligence in

Left: elk, Alces alces. *The massive antlers of elk and other deer are valuable defensive weapons against wolves and other predators.*

Far left: black rhinoceros (Diceros bicornis). *Extremely aggressive and temperamental, black rhinos even use their horns against Landrovers and sometimes put safari parties to flight.*

Left: western diamondback rattlesnake (Crotalus atrox). *Rattlesnakes use their erectile venomous fangs mainly for catching prey, but also in defence.*

Below left: Sahara scorpion (Androctonus australis). *Scorpions seize and tear apart small prey with their pincers. The venomous sting only comes into action with large prey. It is basically intended for defence, never spontaneously to attack humans.*

Below centre: Galapagos giant tortoise (Testudo elephantopus). *Tortoises rely on the passive protection of their carapace.*

Below: Canadian porcupine (Erethizon dorsatum). *The strategy of hedgehogs and porcupines is intermediate between purely passive and active defence. Porcupines can seriously injure an attacker.*

apes, dolphins and humans. We may be far superior to animals in intellect; yet insects, with a brain roughly a millionth the size of ours, have endured almost since time began.

Left: frog (Mantella auriantiaca). *Frogs secrete poisonous skin mucus. Many other animals, such as insects, protect themselves by chemical means.*

TRICKS AND INTELLIGENCE

Deception plays a key role in the life of many animals. Problems that cannot be overcome directly by the animal have been solved by the evolution of many different forms of trickery. In the struggle for survival, these behaviour patterns are economic as they save energy. For example, rather than fighting or fleeing from an opponent, an animal may 'freeze' and camouflage itself against its natural background to go unnoticed.

Intelligence is also a form of economy. Instead of tackling a new, potentially dangerous situation recklessly, an intelligent animal uses forethought. Careful evaluation of all factors enables it to find the most suitable form of behaviour and thus to avoid mistakes.

Nevertheless, in the animal kingdom there are many actions which appear clever but which cannot be ascribed to intelligence. Rigid, inborn behaviour may often appear intelligent because it is so well adapted to the life of the animal. Only when the animal is removed from its normal environment and its reactions tested in an unfamiliar situation can one test whether or not it is really showing true intelligence and forethought in such a situation.

Left: The fox leaves the trap with the bait. The fox's proverbial cunning is in some measure an example of true intelligence. Very adaptable in their behaviour, foxes have begun to enter towns and still roam wild in some city areas of heath and woodland.

Right: in a laboratory experiment the blue triggerfish removes the glass lid to reach a sea-urchin. Does this denote intelligence? It is rather doubtful, because if the container is removed the fish no longer recognizes the connection between the lid and the prey. Even under natural conditions these fish remove obstacles. So this is an example of a largely inborn behaviour pattern.

Right: lying upside down with open mouth, this grass snake feigns death. The purpose of such behaviour is not yet clear.

Below: this plover behaves in a similar manner. It feigns a broken wing and flutters as though helpless, so luring an intruder away from its nest.

Below: like most amphibians, this attractive frog (Mantella gorana) gets some protection from its poisonous skin secretions (left). The death-like pose (right) is a last attempt to bluff an enemy and thus to gain time.

IN EMERGENCY

Most wild animals characteristically show caution and mistrust when confronted with anything suspicious or out of the ordinary. A hunted animal, for example, will often avoid making a movement before a predator has spotted it. Should its precautions prove fruitless and the predator approach closer, thus posing real danger, the animal will react in the only way it knows, though what happens has little to do with intelligence. In such an emergency, there is no time for a carefully considered course of action.

The most obvious defence strategy is to escape at any price. Speed is an obvious asset and many animals, such as antelopes, can outstrip most pursuers. But others have to resort to stratagems that will baffle and outwit their enemy. Indeed, some animals can manage to escape even after they have been seized by a predator. Certain lizards, for example, can shed their tail, which thrashes around when severed from the body, distracting the predator. This ruse gives the lizard a split second chance of escape. Its tail later regenerates, apparently with no ill effects, and thus the lizard survives. This kind of behaviour, however, is not restricted to lizards. Certain spiders, crane-flies and crustaceans shed their limbs as soon as they are seized by a hunter. In an emergency, dormice and other rodents allow the skin of their tail to be stripped off, while Smith's red hare of South America

leaves part of its fur in the predator's mouth.

When confronted by its enemy, an animal's last defence may be to 'play possum' and feign death. Although this does not give the animal absolute immunity because predators will also eat carrion, hunters using sight rather than smell react to movements of their prey and therefore will not respond as much to something that is inert and dead-looking. There are numerous examples of this means of defence in the animal kingdom, the most famous being that of the North American Virginia opossum which, when frightened, transforms itself from an alert arboreal mammal to a convincing twisted 'corpse', thus saving its life.

A stone-curlew guards its nest against a green monkey. As ground nesters, these curlews are particularly threatened by egg robbers. The intelligent monkey has not been deceived by camouflage nor diverted by the desperate guard manoeuvres of the parent bird. When threatened, newly hatched chicks are carried to safety by the parents and when two days old they can themselves escape on foot.

THEFT

Theft appears as a widespread practice in the animal kingdom and can be regarded as a kind of non-violent predation on other species. It is prevalent in many groups; for example among the ants there are very many slave-making species. The principle is that one species of ant steals the pupae of another species which, as they develop into adults, serve as an extra work force. Some ants of the genus *Polygerus* have mouth-parts that have become totally adapted to stealing other ants, so much so that they depend on the so-called slave species to feed them and carry out all the duties of the nest.

Among birds, skuas are experts at thieving the food catches of other seabirds. Skuas are well adapted for the chase as they are strong fliers and have long hooked bills for snatching the food in mid-air. The common Arctic skua, also known as the parasitic jaeger, will often threaten terns, harrying one bird after another to give up its catch.

Close to the nest of a greater black-backed gull, a jackdaw has found a dead rabbit and begins to pluck out the eyes. Given its relatively weak bill, this is the jackdaw's best point of attack.

Shortly afterwards, the jack-daw is chased away by the gull, which uses its more powerful bill to tear the corpse apart.

This young baboon has found a precious prize, an ostrich egg. Grasping the egg firmly, the baboon makes off with the booty. Any confrontation with the adult bird would be extremely dangerous. Eggs contain everything needed for the development of a bird embryo: nutriment, fluid, vitamins. This concentration of nourishing ingredients makes it a valuable source of food for other animals. So eggs have been stolen for as long as they have been laid. Egg robbers are found among insects and also in reptiles, birds and mammals. Naturally, larger species mainly go for bird and reptile eggs.

In the meantime, the jackdaw has removed an egg from the gull's nest, cracks it open and begins to devour the contents.

After a while, the jackdaw is again disturbed. The same gull has found the intruder again and chases it off.

The gull does not recognize its own egg because it is no longer in the nest and is also broken. It eats the contents.

A MURDERER AND ITS EXECUTIONER

Lightning-fast agility triumphs over brute strength: a drama in five pictures.

1. The small wasp Methocha *allows itself to be seized by the giant tiger beetle larva. Thanks to its slenderness, it is not crushed by the beetle's mandibles.*

2. The wasp has freed itself from the lethal embrace and paralyzes the larva by a sting in the thorax.

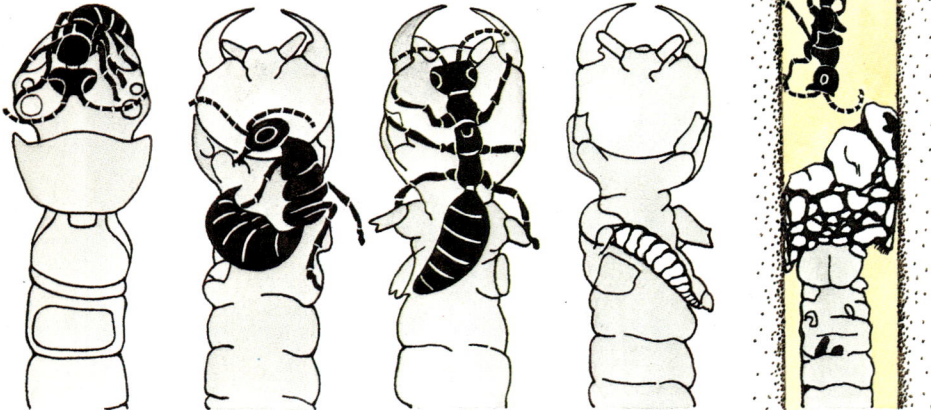

3. An egg is laid on the paralyzed predator.

4. In due course, this develops into a wasp larva which slowly feeds on the paralyzed but still living beetle larva.

5. After it has laid its egg, the wasp closes the burrow entrance with sand and small stones.

Facing page: the predatory larva of the tiger beetle lives almost completely hidden in its self-dug sand burrow. Only the upper part of the head with its small eyes is visible. When an insect approaches, it leaps out and seizes it with the powerful mandibles.

The field tiger beetle is a medium-sized metallic green beetle with white spots, found in sandy areas. It locates its prey with well-developed compound eyes and overpowers it in a lightning-fast dash. However, we are not concerned here with this masterful predator, but with its larva. The juvenile form of the field tiger beetle (*Cicindela campestris*) is for some insects just as dangerous as the adult beetle. The larva does not pursue its prey, but lies in wait for it. The whole body of the larva is hidden in a burrow which it has dug for itself in the sand. Only the upper part of the head, with its six simple eyes, and the neck are visible from above. These parts seal off the burrow and blend in with the surrounding sand. As soon as an unwary crawling insect comes close, the larva rushes out of its burrow and seizes the prey with its powerful dagger-like jaws.

Once it has been seized there is no escape for the unfortunate victim; as often as not, it is immediately decapitated.

In the insect world there are few predators, however awesome, that do not themselves have a deadly enemy. In the case of the tiger beetle larva, for example, the foe and conqueror is a tiny wasp, *Methocha*. This wasp actually allows itself to be seized by the powerful fangs of the larva, but thanks to its slender build and agility, it easily manages to break free. The wasp then paralyses the beetle larva with a rapid sting, lays an egg on its body and inters the larva in its own burrow. After some time the wasp egg hatches into a larva which feeds on the flesh of the paralysed beetle larva.

Solitary wasps such as *Methocha* show many fascinating examples of the sheer power of instinctive behaviour over an adversary much more formidable than themselves. As adults, hunting wasps feed peacefully on the nectar of flowers, but as the time approaches for egg-laying, the female wasps seek out prey on which to deposit their eggs. The prey is merely paralysed so that it remains fresh for the larva once it hatches out from the egg. Each kind of hunting wasp tackles different prey, ranging from grasshoppers and bush crickets to caterpillars and large spiders.

ANGLER AND MARKSMAN

Angling is a form of hunting that is by and large a matter of deception. Whether or not the trout bites depends primarily on the quality of the proffered prey or bait. Rather than pursue the fish, the angler exploits the senses and hunting instincts of his victim, thus saving effort and energy. But the trout would long ago have become extinct had it been unable to combat these tactics. During the course of its evolutionary history, the fish has developed increasingly efficient sense organs and a brain in order to distinguish between true prey and a bait. The angler, in turn, would similarly have gone hungry had he not managed to refine a bait which was at first rather crude. The age-old running battle between the angler and his catch has enabled both to survive and to develop even further. There would be no point in the angler trying to devise more effective types of bait which might seriously threaten the population of his chosen victims and thus jeopardize his own future.

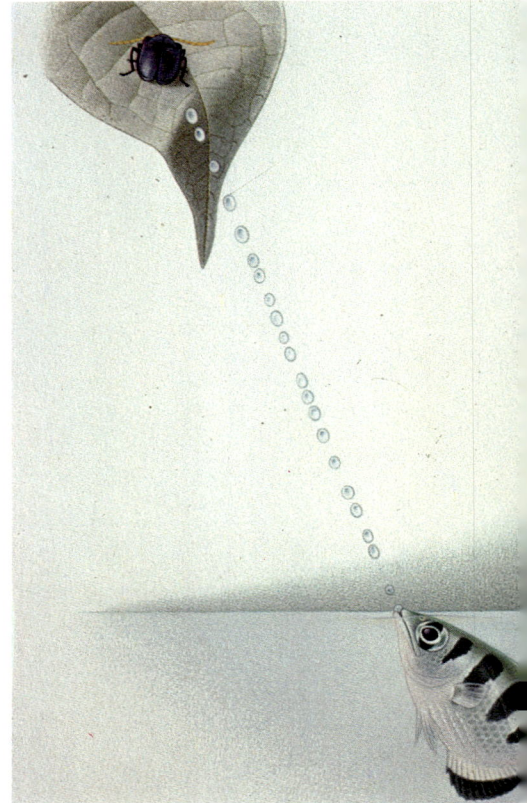

Although humans have developed an astonishing expertise in the field of ballistics, they are by no means the first marksmen. Shooting as a means of defence and for capturing prey is found in the coelenterates, a relatively primitive animal group that includes

Above: the bombardier beetle (Brachinus) has two abdominal storage chambers secreting two chemical substances which, separately, are relatively harmless. Both chambers have exit ducts leading to a third chamber where the two components react to produce an explosive jet, at more than 100°C, which is squirted at enemies. The mechanism resembles the shooting of a rocket.

Centre: the archer fish (Toxotes) can shoot down insects above water level. Adults about 20 cm long can shoot a distance of up to 1.5 m. A fish, lurking just below the surface, sights prey, focuses on it with both eyes, and takes up a position as near vertical as possible in order to reduce the refraction of light at the water surface. By suddenly closing the gills, it expels a stream of water from between the tongue and the roof of the mouth. The shot knocks the prey from its perch and it falls into the water.

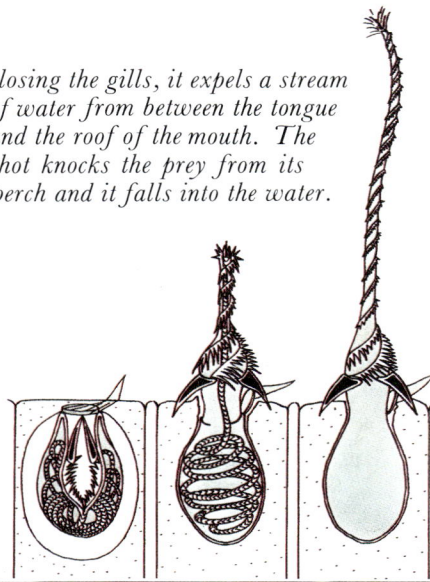

Left: the sting cells of sea-anemones, corals and jellyfish represent, from the evolutionary viewpoint, a very ancient form of shooting. At a touch from an outsider, the tiny cell explodes (left) and shoots out a harpoon-like projectile which pierces the skin of the victim and anchors itself by hooks. This shot can be lethal to invertebrate animals. Large jellyfishes and sea-anemones with thousands of sting cells can cause severe injury, sometimes even to humans.

jellyfishes and sea-anemones. These are soft-bodied organisms without true sense organs. They probably owe their survival in large measure to their oval or cigar-shaped stinging cells which,

if touched, shoot out a filamentous projectile, armed with barbs and poison.

The larger jellyfish and their relatives such as the Portuguese-man-o'-war, have enormous batteries of stinging cells on their long tentacles, which are used to anchor as well as to stun the prey; by contraction of the tentacles, the latter is hauled up towards the mouth region where it is digested.

Accurate marksmen among spiders include the bola spiders which actually lasso small moths with a thread containing a sticky droplet, perhaps with an attractive odour, at one end.

Below: Some deep-sea angler-fishes have a bait with a light organ attached to their body. Animals which are attracted by the light disappear between the jaws of the predator.

25

TRAPS

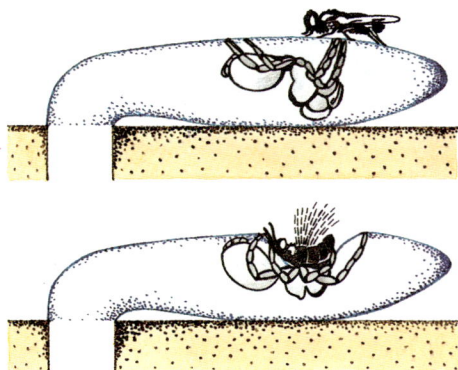

Right: Web spiders are excellent trappers. Orb-web spiders such as this zebra spider (Argiope) usually spin a new web every morning. For this they choose places where the flight traffic of their prey is particularly dense.

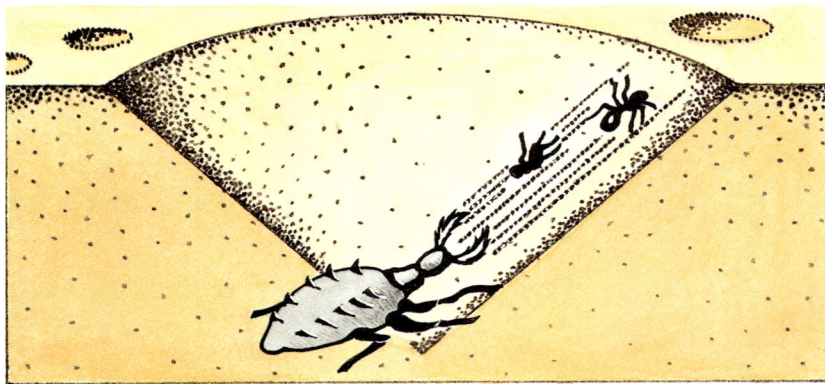

Above: the larvae of ant-lions prey on ants and other insects that fall into their funnel.

Right: the spider Masto-phora entices male moths by letting a drop of glue whirl round from one leg and at the same time secretes a sexual hormone attractant. The glue drops on to approaching victims.

Bottom: spiders of the genus Atypus build silken tubes which extend along the ground like fingers of a glove. When a small insect accidentally runs over the tube, it is seized from within, dragged in and eaten. The hole is later repaired.

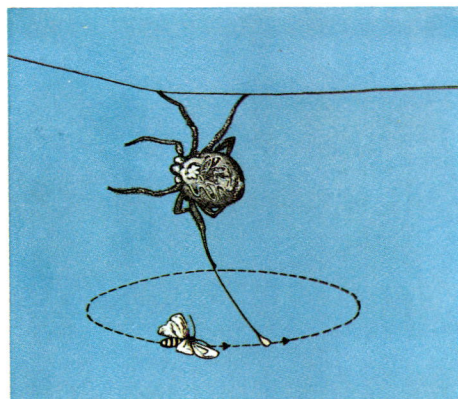

Traps are used by a number of poikilothermal (so-called cold-blooded) animals which expend little energy in maintaining their metabolism and are therefore able to wait in patient inactivity for long periods of time. This art has been particularly well developed among insects and spiders.

The ant-lion, a larval insect, builds its funnel-shaped pit in a well protected area of loose, dry sand. It moves backwards, using its head to toss particles of sand outwards until it has dug out a pit. Under favourable conditions, this process lasts fifteen to twenty minutes. The larva then positions itself obliquely at the base of the funnel, its large suctorial fangs spread out, and waits. Any insect, such as an ant, which accidentally falls into the slippery pit is seized by the powerful, curved jaws and its body contents sucked out. Sometimes the prey tries to escape, whereupon the ant-lion flicks sand over it, so that it slides back down the slope to the bottom of the pit and is then devoured.

The most perfect setters of traps are spiders, although not all of them spin a web. The trapdoor spider digs a deep burrow in the ground using its fangs to loosen and remove the soil, lining the shaft with silk and the entrance with a lid of silk and soil. When an insect approaches, the spider opens the lid, rushes out and seizes the prey. Nor do all the web-spinning spiders construct a typical radial web. The spider *Dinopis*,

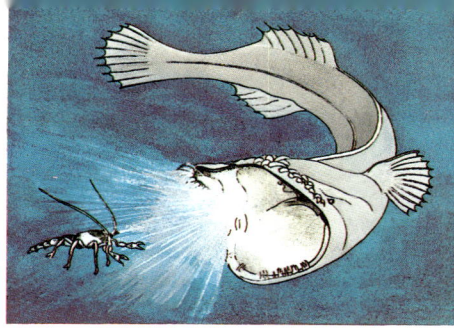

Left: in the black ocean depths, light is an unaccustomed phenomenon. Deep-sea anglerfishes lure their prey into the mouth by a luminous bait hanging from the palate.

for example, which hunts at night, holds its web stretched out between the front legs. With its large eyes it locates the prey and throws the web over it, like a cast-net. By jerking its body backwards it ensures that the victim is hopelessly entangled in the sticky threads. There are dozens of other types of web spiders, each with its own special refinements. The traps are effective enough to procure a regular food supply.

Above: these trapdoor spiders (Ctenizidae) live in earth burrows and close the entrance with a spun lid. At night they lie in wait behind the entrance, stretching out the front legs as feelers. When these signal prey, the spider leaps out and catches the victim.

Left: sea-anemones use their stinging arms for catching food. The victim here is an atherine.

THE GALAPAGOS FINCHES

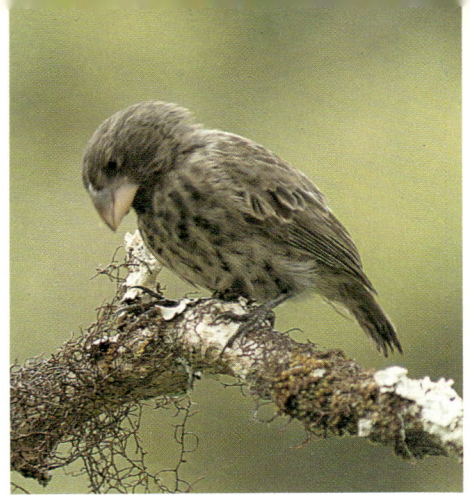

Because there are no food competitors on the Galapagos Islands, the original seed-eating finches have exploited a variety of food sources. While one exclusively plant-eating species remains (right), other species take a mixed diet or have come to specialize in feeding only on insects. One long-billed Galapagos finch (below) has taken over the role of the woodpeckers. They show astonishing dexterity, extracting insects from burrows in the tree-bark with a twig.

The most common land birds of the Galapagos Islands are dull-coloured sparrow-sized finches. Early in his stay on these islands, Charles Darwin remarked that these birds had different sized and shaped beaks. He studied their diet and behaviour and they became the models of his famous theory of evolution.

Darwin postulated that only a few song birds had ever reached the Galapagos from the South American mainland a thousand or so kilometres away. These first settlers were finches, probably of a single species. Their normal food most likely consisted of plant seeds, which were in limited supply. There was, however, a surplus of berries, plant shoots, insects and spiders, because there were no other small birds to feed on them.

Searching for food, the Galapagos finch looks into the worm holes of insect larvae until it discovers one.

So, unhampered by food competitors, the finches expanded their diet and over the course of thousands of years, became specialized to tackle a diverse range of

Now it tries to reach the larva with its beak. For a finch it has a relatively long beak, but often this is not sufficient.

It cannot break into sound timber and, unlike a woodpecker, it does not have a long "angling" tongue (drawing, left); so it picks up a cactus spine or a small twig (right) and starts to poke into the hole with it.

foods. Today there are thirteen species that probably originated from a common stock. The ground finches comprise six species that feed mainly on seeds and insects, but one species has a long curved beak and feeds on cactus flowers. A second group, the tree finches, have large parrot-like beaks and there are also two species which, like woodpeckers, feed on insects taken from tree trunks and branches. Lacking the strong, sharp beaks of woodpeckers, they use twigs and cactus spines as tools to extract insects from holes.

Usually the larva tries to leave the hole. At this moment it is seized.

Prey in beak, the finch hops away to eat it in a convenient place.

ANIMALS THAT USE TOOLS

Various animal species know how to use tools. In certain cases, as with chimpanzees, this implies a measure of intelligence, but in others it is undoubtedly an instinctive, inborn activity, as they do not possess the mental equipment

Above: when an Egyptian vulture discovers an unguarded ostrich egg, it picks up a stone and bangs it against the egg until the hard shell cracks.

Right: tailor birds perforate the edges of a large leaf and use plant fibres to stick them together in the form of a cornet, in which they build their nest.

necessary to think things out in advance. Sea otters dive for clams and usually bring up a stone as well when they return to the surface. Swimming on their back, they place the stone on their breast and strike the clam against it until the shell breaks. They do the same with sea-urchins. Some bowerbirds in New Guinea squash berries and use a stick or a piece of bark as a brush to paint their bowers with the coloured sap; and we have already mentioned the two species of Galapagos finch that use twigs to fish insect larvae out of their burrows. All these are examples of something more than purely instinctive behaviour.

Facing page, right: the weaver ants of Africa and south-east Asia have their colonies in treetops, and make their nests from leaves. Several ants join in drawing the leaves close together and binding the edges. They use their own larvae as shuttles, as these contain the necessary silk glands.

Left: chimpanzees, the most advanced of all animals, use tools and weapons. In their search for food, they drum with sticks on hollow tree-trunks, poke straws or twigs into termite holes, and then eat the insects which have seized hold of them. They sometimes prepare the twigs for this purpose, virtually making tools, an ability once thought to be possessed only by humans. They also threaten their opponents with sticks. The chimpanzees of certain populations attack and batter enemies with branches up to 2 m long. Swiss zoologists in West Africa have recently discovered that these anthropoid apes crack nuts with hard objects.

ANTS THAT GROW FUNGI

Using its razor-sharp mandibles, a leafcutter worker ant has detached a piece of leaf. Since thousands of ants participate in this work practically every leaf in a plantation may be destroyed within a very short period. The sharp-pointed mandibles of the ants are also used in defence, and those of larger workers are even capable of penetrating human skin.

regions of America. With over a million worker ants, their colonies are among the largest aggregations of ants to be found anywhere. They build nests in the soil that are several cubic metres in volume, consisting of thousands of chambers, with innumerable entrances and exits. Even the much-feared army ants will not venture into such a colony.

In the course of an existence of over twenty years, a colony of leafcutter ants moves several tons of foliage underground. The gigantic biomass is not eaten but is used as a medium on which to grow fungi. In special chambers of the nest the leaves are chopped up, chewed into paste and fertilized with faeces, until they become an

In parts of South America, travellers have been astounded to see what was once a luxuriant tropical plantation turned overnight into an area of devastation, with only bare branches rising up into the air. It is as though the ground had swallowed up all the green foliage and, in a sense, this is exactly what has happened.

In a single night, armies of leafcutter ants can carry the vegetation of an entire plantation down into their nest. Each individual ant rips off a leaf or section of leaf with its powerful jaws and staggers back with the heavy load to the anthill. Then it repeats the operation as often as may be necessary. The leafcutter ants live exclusively in the tropical and sub-tropical

Left: view inside the nest of a colony of leafcutter ants. The fungus cultures can be recognized as diffuse white masses which are being tended by the small "in-service" ants. Here too are the ant pupae in which the future eyes and antennae are discernible.

ideal nutritional substrate for one particular species of fungus. Substances secreted by special glands in the thorax of the ants inhibit the growth of unwanted fungi and bacteria. Careful tending of the network of fungal filaments encourages the formation of so-called ambrosia heads, tiny white bodies which serve as food for the ants. When a mature queen ant breaks out to establish a new nest, she takes with her, in a body pouch, a few fungal filaments from the parent colony. Later, at the place where she lays eggs, she disgorges this little store and fertilizes it with faeces. This manuring process starts the fungus growing and a new fungus garden is established. This will serve as food for the entire new ant colony.

Below: the pieces of leaf are transported in long columns into the nest where, by being chewed and mixed with saliva, they are prepared as compost for the fungus cultures. The fungus gardens, carefully tended by tiny "in-service" ants, supply all the nutritional requirements of the colony.

The small Indo-Pacific boxer crab (Lybia tesselata) confronts its enemies with two sea-anemones which it carries around with the pincers of its front legs. The pincers have small denticles and hold the poisonous sea-anemones so firmly that some force is needed to prise them off. The crab uses the sea-anemones not only as weapons, but also as collaborators when seeking food, commandeering some of the victims caught by the sting cells. In this partnership, therefore, the crab profits considerably more than the sea-anemones; the latter, however, have the opportunity of trapping more food animals by being constantly moved around. A hermit-crab (Diogenes edwardsi) behaves in a similar way to the boxer crab. It carries a sea-anemone on its left pincer and uses it to close the entrance to its home.

MIMICRY

A bumblebee visits a flower (*left*). Bumblebees are friendly but well armed, and can use their sting if necessary. On the other hand, this clearwing (*right*) is a completely harmless moth which is protected from enemies by looking like a bumblebee.

The black-and-yellow of this wasp (*left*) implies that it is a dangerous antagonist. It has a number of mimics from various insect groups, e.g. the wasp beetle (*right*), the pupae of the common magpie moth (*facing page, left*) and the wasp-spider (*facing page, right*).

Ants (*left*) destroy a large wood-wasp (Urocerus gigas). Many birds avoid ants, so it is an advantage for other insects to look like them. In fact, there are ant-like spiders and bugs. The picture (*right*) shows an ant-bug (Myrmecoris).

The hornet (*left*) is the largest social wasp in Europe. Its sting is so feared that even large animals leave it alone. The completely harmless hornet-clearwing moth (*right*) takes advantage of this. Its similarity is so striking as to deceive anyone but an expert.

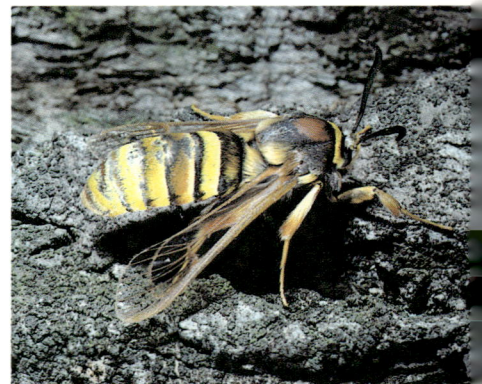

Around 1860, the English naturalist William Bates noticed that among a number of heliconiid butterflies he had caught in the rain forests of Brazil there were some which, although they appeared to be heliconiids, were definitely not so. In fact, they were identified as members of the cabbage white family. Yet the similarity in appearance was so striking that, in

nowadays recognized. All the illustrations shown on this double spread are examples of Batesian mimicry, in which one or more harmless species adopt the appearance of a poisonous species. In addition, there is Mullerian mimicry in which several dangerous species resemble one another, thus teaching predators that this body pattern means 'poison'.

Below: the death's-head hawk-moth is a real virtuoso of mimicry, in three different ways. Its death's-head pattern imitates the face of a vertebrate; a form of protection primarily against birds. At the same time, its abdomen bears the pattern of wasps and hornets, and its size gives it a resemblance to a giant hornet, a point surely taken by this hungry garden dormouse. However, this apparel is of no use to the death's-head when it enters a beehive to suck honey. To avoid being killed by the bees, the moth imitates their sounds and this keeps them quiet: a case of acoustic mimicry.

Bates's opinion, it could not have been accidental.

What then could be the basis for this remarkable likeness? Bates reasoned that heliconiids are poisonous and are therefore not eaten by birds. So if a non-poisonous butterfly, such as a white were to pose as a heliconiid, it would presumably be avoided by birds, provided both were sufficiently similar and the deception was not noticed. The warning coloration of the poisonous species would thus become a protection for the innocent but fraudulent species. This is the principle of Batesian mimicry, named after the naturalist.

Several types of mimicry are

LEARNING GAMES

Below: the lion cub plays with its mother, just as later it will bring down a gnu or antelope. Lions and lionesses are customarily very tolerant of their offspring.

Learning is an old and indispensable recipe for success in the natural world. As they learn, animals become capable of adapting to new situations which they might not be able to deal with by relying on instinctive behaviour programmes dictated by heredity. Such situations may include environmental factors that change rapidly and unexpectedly, posing a threat to the animal's survival. Even insects, which are renowned

Above right: a small brown bear in Basel Zoo plays at seizing a tree trunk which its sibling tries to defend. Playful bear cubs are extremely endearing to zoo visitors. Later, during territorial fights, things become much rougher.

Right: young mountain gorillas wrestle among the branches of an African forest. The anthropoid apes have excellent learning ability. The lessons they learn from play are also useful, giving them complete mastery of the three-dimensional world around them, and an understanding of their fellow apes.

for their fixed inborn behaviour patterns, can, to a certain extent, learn. By and large, however, the crucial importance of the capacity to learn is most markedly evident among mammals. A lion cub does not know how to hunt as soon as it is born; it has to be taught to do this by its parents. Young gorillas have only a few inborn capabilities at birth. Almost all behaviour patterns – building a shelter, finding food and generally coping with life's problems – must be learned.

Young animals evidently enjoy learning and often do so through

The domestic cat depends entirely upon its mobility and dexterity. Adult cats are solitary and refrain from tussling with another cat because play might develop into something more serious. Playful kittens therefore use the freedom of their age to fool about in preparation for later life.

Large-toothed hyraxes are ideal victims, being hunted by numerous predators such as leopards, eagles, large owls and hyaenas. They can only survive by constant vigilance and readiness to flee. This is reflected in play, in which chasing one another is an important activity. As the picture shows, even youngsters of different hyrax species will play together.

the medium of play. By kicking and chasing a ball of wool, a kitten is developing a behaviour pattern that it will later use more profitably to catch a mouse. The apparently playful antics of wolf cubs are a foretaste of later fights for dominance with rivals. On the other hand, small and frequently threatened animals, such as squirrels, routinely practise escape as a form of play.

It is surely no accident that the young of the most intelligent animals, namely apes, dolphins and carnivores, are among the most playful of all. With their highly developed brains, they have a very great capacity to learn, and they have considerable ability to adapt innocent play experience to dealing with new situations and the real problems of everyday life. After hundreds of years in which rigid, dull and laborious methods of learning have been the rule, we have only recently come to realize the advantages of play in the school curriculum. Children who learn to understand through games, perhaps using computers, prove the value of such methods.

39

THE WAY ANIMALS CAN LEARN

Sea-lions have by nature a marked sense of balance. This is probably related to their swimming habits. An animal trainer has only to encourage this talent. He is helped in this by the playfulness of the animals.

Education of a squirrel. An inexperienced squirrel recognizes by instinct that a nut is food and knows that it has to be opened; but it has to learn how to do this.

For many years zoologists have argued about whether animal behaviour is inborn or whether it has to be learned. Today most research workers in this field are of the opinion that there is a precise interaction between heredity and experience. It often happens that individual elements of behaviour are inborn, but the young animal still has to learn by trial and error how these elements can be combined in a sensible way.

As an example, consider how squirrels open nuts. An experienced animal opens a nut by gnawing a longitudinal groove and a hole at the top. It then inserts its lower incisor teeth into the hole and with a dexterous twist splits the nut into two halves. In this process inborn elements probably stimulate the action of gnawing as well as the movement necessary to split the nut. Inexperienced squirrels, therefore, often gnaw the

Phase 1: the nut has certainly been opened, but with too much effort, as shown by the large amount of gnawing.
Phases 2 and 3: the squirrel has made progress; it now starts to concentrate on the most suitable places.
Phase 4: the trick has been learned; the nut has been gnawed along the protruding edge.

nut quite aimlessly until they learn how to gnaw a groove parallel to the run of the fibres and then to use the inborn splitting movement.

Not all animals are capable of learning equally well. The extent to which an animal can learn is probably established genetically. It would scarcely be likely for a pigeon to open a nut or for a horse to dig a hole. The timing of the learning process also varies. In the

lifetime of many species there are periods when an animal is particularly able to deal with certain situations. When all the conditions are right, animals are capable of the most astounding feats. Rats can find their way through a complicated maze within a very short time and such tasks are often performed, as a matter of routine, in their natural environment. Song birds have an excellent acoustic

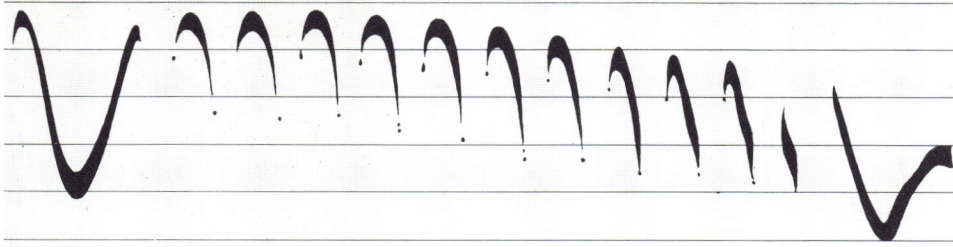

sense. After a period of months they can still differentiate between two notes which are only half a tone apart.

In the case of circus animals – performing dogs, horses, seals and the like – what we see and applaud is the result of learning. The animals have particular inborn abilities which the trainers develop, and use in order to teach them tricks.

Above: Rolf Knie junior with his horse Pellarin in a difficult training routine. Both lie down beneath a woollen blanket. Pellarin has to learn to remove the blanket from Rolf.

Left: Pellarin drags off his saddle, another difficult training procedure, which involves apparently unnatural tricks based on the animal's inborn abilities. The art of the animal trainer is to use and exploit these gifts.

41

BROOD PARASITES

cuckoo's eggs and rear her young. The eggs of the cuckoo are remarkably similar in colour and pattern to those of the host bird. The cuckoo cannot, however, vary its

Each of the fifteen different whydah species parasitizes a particular finch species. The finches recognize the characteristic calls, the begging movements and, above all, the striking and specific throat patterns of the whydah chicks. The young parasites are astonishingly similar to the young of their host.

Above left: paradise whydah with its two-week-old chick; alongside is its host, a melba finch, with its chick of the same age.

Above right: atlas whydah with its young and (far right) a firefinch with young.

Right: Fischer's whydah with three-day-old chick. Far right: a purple grenadier with a nestling of the same age. In all cases the hosts rear the baby whydahs with their own chicks.

The European cuckoo is indisputably the best-known example and most effective practitioner of brood parasitism, the habit of depositing eggs in the nest of another species. It has become a real specialist, leaving the troublesome job of rearing its offspring to other birds.

The female cuckoo lays its eggs singly in the nests of song birds. Although the latter are five to six times smaller than the parasite, they promptly incubate the

egg type at will. An individual female is adapted to parasitize one or a few host species which lay similar eggs. Thus, there are cuckoos which specialize in parasitizing white wagtails, robins, hedge sparrows, reed warblers and so on. Their offspring will in due course choose the same host species.

The young cuckoo grows incredibly fast. After three weeks it is about fifty times heavier than it was at the time of hatching. The foster-parents become quite thin from the effort of collecting so much extra food for their un-

Left: a wagtail feeding a young cuckoo, which has already left the nest. The cuckoo eggs match in colour and size those of various song-bird eggs. Above are the eggs of a blackcap and of a cuckoo.

welcome guest. The gaping orange mouth of the young cuckoo triggers an instinctive response. Nor will the stranger, with its voracious appetite, tolerate competitors in the nest. When it is a mere ten hours old, still naked, blind and apparently helpless, it starts, using the rear part of its body, to push all the eggs or even the other hatched chicks out of the nest. Soon it fills the entire nest, and the foster parents have to mount its back to feed it. Although the young cuckoo is devotedly fed by its foster parents, it lives a dangerous life. About three out of four young cuckoos fall victim to predatory animals or meet with an accident before they are fledged. In order to make up for these losses, the female, during each season, lays about ten to twelve eggs in as many nests.

There are some 140 different cuckoo species in the world, of which 65 per cent tend their own offspring. Some ducks lay their eggs in the nests of other ducks.

This row shows, left to right, the following eggs: red-backed shrike and cuckoo, chiffchaff and cuckoo, whitethroat and cuckoo, garden warbler and cuckoo, reed warbler and cuckoo.

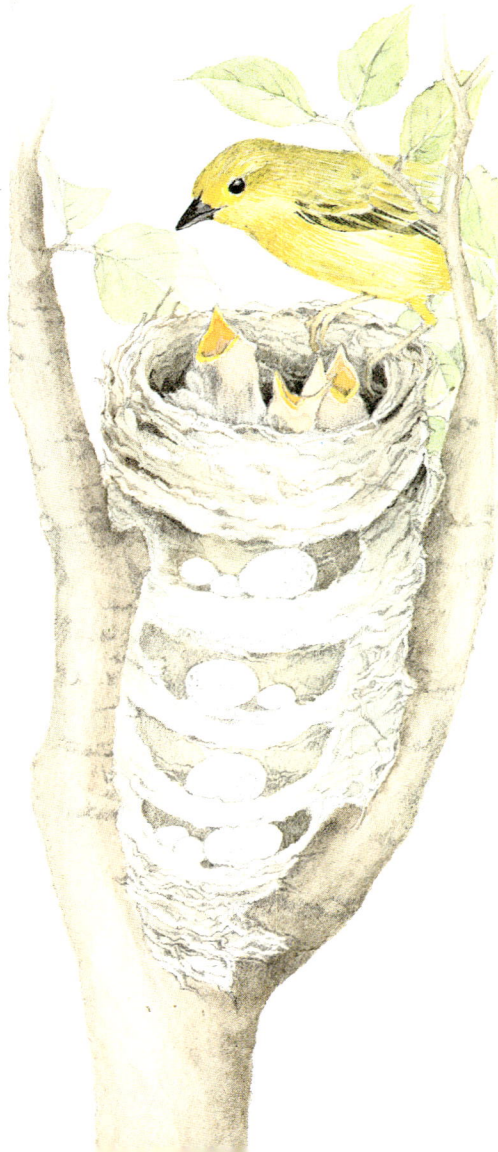

Left: the eggs of the black cowbird of North America have been found in nests of over 200 song-bird species. Many species accept the foreign eggs, but the yellow warbler (Dendroica petechia) notices the deception, lays nest material over the clutch and starts to lay afresh. This performance is repeated several times until finally there are no more parasite eggs in the nest.

43

THE GENIUSES

Dolphins are extremely intelligent animals. Their brain is larger than that of man, and the cerebral cortex has at least as many convolutions.

Many astute patterns of behaviour observed in animals appear intelligent to us. Yet often this is not the case. True intelligence is shown when an animal is presented with a completely new problem. If the

Japanese macaques (Macaca fusca) *have the most northerly distribution range of all monkeys. In winter their natural environment is covered with snow and ice. Perhaps it is these harsh conditions that have made these animals so capable of learning and adapting. They have to be inventive in order to survive.*

animal is genuinely intelligent, it will be able to deal with such situations. On the other hand, if it merely possesses instinctive cunning, its behaviour will remain inflexible and it will be unable to adapt.

The difference is shown particularly clearly in the use of tools.

When a Galapagos finch uses a stick to probe for insects, it is displaying an inborn pattern of behaviour. It would not be able to change this behaviour if other implements were suddenly required for obtaining food. It is quite different in the case of a chimpanzee. They too utilize tools, fishing for ants and termites with the aid of stalks and twigs. However, in contrast to the finch, they have first to learn this type of behaviour; and they are capable of using other tools should the situation require it. Chimpanzees in the Yerkes research centre in the USA have even learned to obtain their food from an automatic machine by using a coin.

Intelligence leads to flexibility by showing insight into a situation. In general, intelligence always occurs in animals where the brain has reached a certain degree of complexity. For example, an octopus is capable of carrying out fairly complicated manoeuvres in order to get at its prey. Dolphins, responding well to captivity, can be taught to perform various tricks and tasks, possessing an exceptionally complex brain.

Innumerable intelligence tests have been carried out in the laboratory with chimpanzees, orangutans and other species. Yet even in their natural environment, as scientists have discovered, apes and monkeys can develop new types of behaviour which will be passed on to later generations.

Japanese scientists have conducted many experiments on the

individual and group behaviour of Japanese macaques in their natural island surroundings. Behaviour patterns were observed to differ considerably from troop to troop, and even among individual monkeys. The researchers noticed that a number of young

macaques were in the habit of washing dirty food, such as potatoes, in water. Other macaques sat watching and eventually began to do the same. Evidently they had all discovered that potatoes cleaned in sea water were more tasty. This new pattern of behaviour, which had not previously been noted among the members of the troop, subsequently became part of their behavioural repertoire. It is interesting that only juveniles and females proved capable of learning. The old males never accepted this new advance. Yet the offspring acquired the new technique when they were young members of the troop.

Above: a young female macaque has learned that dirty potatoes can be washed. She has subsequently discovered that if cleaned in sea water, the food will be tastier.

Left: despite its intellectual capacity, the orang-utan has proved less adaptable when confronted by humans and is now endangered.

Following double spread: a typical behaviour pattern of Japanese macaques is to bathe in hot springs during winter.

SURVIVAL
STRATEGIES

The relationship between predator and prey in the animal world is a subject that exerts much fascination, deserving careful study yet arousing powerful feelings. Age-old fears and prejudices make it more difficult for us even today to express an unbiased opinion on this topic. As former victims of large predators, we tend automatically to side with the weaker party. Yet we were also hunters ourselves, and until quite recently we had few compunctions about killing wild flesh-eating animals, regarded as fair game. If we argued, in justification, that this would benefit the prey, we were mistaken. One result of this large-scale hunting has been an increase in the population of plant-eaters, which menace the world's vegetation. So there is no need for us to protect prey animals from their predators. Over a very long period of time, the hunted have learned by experience to look after themselves.

In this chapter, we will consider, from the prey's viewpoint, the never-ending conflict between the voracious appetites of the larger animals and the survival instincts of the smaller species, bearing in mind that we have the power to upset the balance. In a way it would be true to say that predator and prey are not only enemies but also partners. The pressure they exert on one another stimulates further evolutionary development.

Left: prey animals have several different methods of dealing with predators. The first stage is camouflage: it is undoubtedly best not to be discovered. However, if this happens there remains bluff, flight and defence. Even in a desperate situation like the one illustrated here, a victim can try to defend itself when it has nothing more to lose.

Right: the dorsal pattern of a death's-head hawk-moth is surely an example of bluff, since it resembles the face of a vertebrate. Such a message is intended primarily for insects' main enemies, birds that hunt visually. Laboratory investigations indicate that birds react very strongly to all kinds of eye patterns.

DECEPTIVE EYES

False eyes are signals, and as such must be conspicuous. On the other hand, a moth has to be inconspicuous. Only when its camouflage is seen through will the scaring effect of false eyes come into operation. Like many moths, the eyed hawk (Smerinthus ocellata) solves this problem in an ingenious way. The eye pattern is found on the upper side of the hind wings; these are normally covered by the camouflaged forewings, which in a split second are drawn back to reveal the "eyes".

Patterns resembling eyes are found on the plumage of birds, the scales of fishes, the shells of tortoises, and, above all, on the wings of butterflies and moths. There are all sorts of subtle variations ranging from simple dark spots to astonishing replicas which even imitate the iridescence of the pupil. The perfection of some of these patterns is quite remarkable, and even rather daunting when, in a casual inspection of some foliage, one is suddenly confronted by two large,

dental. Others believe that they are there for a purpose, to bluff and deceive the enemy, and that the striking pattern diverts the attention of possible predators from the really vulnerable parts of the animal's body. In addition to the large conspicuous eyes on the top of the wings of certain butterflies and moths, there are often dark spots visible along the wing edges. The general consensus of opinion appears to be that such eye-spots are imitations of the eyes

Facing page centre: false eye of a praying mantis.

round, staring eyes. Only from closer observation does it become apparent that this is only a moth, exposing the eye-spots on its hindwings by very rapidly flicking its front wings forward.

Many research workers were and still are of the opinion that these patterns which appear so vividly to resemble eyes are really only a trick of nature, purely acci-

of vertebrate animals. Consequently, they serve principally as warning signals that frighten off birds. A bird which has found a moth will involuntarily recoil when it suddenly sees a pair of eyes, quite motionless, that apparently belong to a large animal, thus providing time for the intended victim to escape. Experiments with false eyes support this theory.

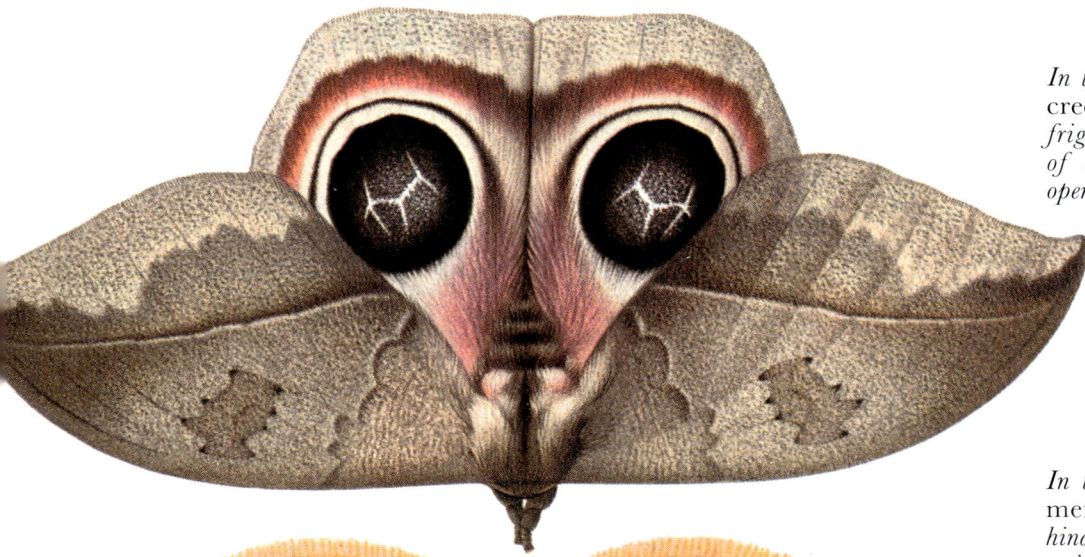

In the African moth Pseudo-creobotra wahlbergi *the fright position appears as a pair of eyes when the wings are opened.*

In the American moth Auto-meris *the eye patterns on the hind wings are particularly striking. In South America there are more than a hundred species in this genus, each with its own characteristic eye pattern. Here, too, the forewings provide camouflage.*

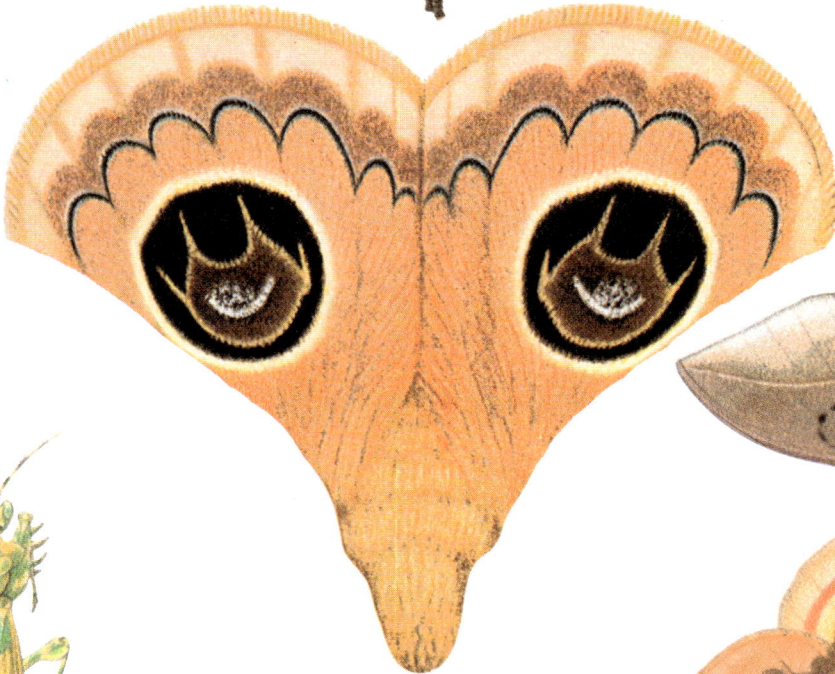

The African moth Lobobunea ammon *performs dancing movements to show off its demon-like false eyes. It repeatedly flutters up from the ground, finally becoming motionless with its hind wings stiffly erected.*

THREATENING GESTURES

The threat repertoire of an African elephant includes extending and flapping the large ears, trumpeting and pretending to charge. These manoeuvres are sufficient to deter the aggressor who can never be sure if they herald a real attack.

The threat behaviour of a leopard reflects a conflict between fear and intimidation. Only opponents that inspire respect are threatened.

Various animal species habitually resort to visible threats in their confrontations with enemies and rivals. Such warnings are usually intelligible to humans. A good example is the threatening behaviour of a large bull elephant about to charge. With its huge ears widely spread, and emitting loud trumpeting noises, the purpose of such behaviour is abundantly clear. Moreover, there will probably be a good reason for such a threat, such as infringement of

territory or fear for the safety of offspring.

In many cases, threatening behaviour is associated with a display of weapons: crabs present their pincers and snakes bend back the front part of their body, as though they were going to strike: some open their mouth wide and show frightening poisonous fangs. Many predators, as well as horses and certain deer, bare their canine teeth, although these are not particularly well developed in the females. This last example shows that threats may often be simply a matter of bluff.

Mammals have a particularly wide range of facial expressions

The threat position of this caterpillar is a bluff: the face is not real.

*Following double spread: the ruffled threat position of this barn owl (*Asio otus*) is enhanced by the circular face and round eyes. Even though there is nothing but air behind the feathers, the attitude inspires respect because owls defend themselves fiercely.*

*Threat position in preparation for striking: this Malayan tree snake (*Ahaetulla*) resembles a coiled spring.*

and body adaptations. They can move ears, mouth, eyes and nostrils, erect their fur, bend the tail in all directions, and distort the entire body into a whole series of impressively intimidating attitudes. The typical threat position of the cats, both wild and domestic, can be interpreted as a blend of attack and defence. The head with the ears laid back flat implies defence, while the rear part of the body with the tail held high and the back sharply arched suggests that the animal is about to attack. It is vital for another animal to interpret such signals correctly and adopt the appropriate and acceptable response.

Left: the principle of making the body appear as large as possible and of showing weapons is adopted by this small South American shore-crab when preparing to flee. Usually, however, it retreats with lightning speed to its burrow.

*Below: the threatening position of this gannet (*Sula bassana*) is no empty gesture. Gannets defend their breeding site very actively, even against larger intruders.*

53

HISSING AND DRUMMING

In the United States and Mexico, rattlesnakes epitomize danger. In the arid regions where they occur, their threatening rattle is one of the most terrifying sounds. The tail rattle consists of interlocking horny segments.

Right: the huge collar of the small Australian frilled lizard (Chlamydosaurus kingii) is an organ designed to impress; it makes its owner appear larger. When fully spread, the collar, supported by cartilage, may measure some 30 cm across, which is more than the reptile's body length. The optical effect is reinforced by a furious snake-like hissing from the throat.

Not all animals are content to restrict their warning behaviour to visual signals. If continuously and persistently harassed, many will call upon a whole repertoire of alternative threats and warnings, including those associated with sounds and smells.

The rattlesnake is a famous example of an animal that relies on acoustic warning signals; and skunks are perhaps the best known animals which emit a foul-smelling fluid to deter enemies. Yet here again, an element of bluff is sometimes involved.

When caught, a grass snake at first flickers its tongue and hisses. It flattens its body by spreading out the ribs, gives sham bites and finally empties the contents of its stink glands. It may follow this up by vomiting its food and, though rarely, inflicting real bites. If they are not then released, some grass snakes will coil up and hide their head. Rolling the eyes, convulsive twitching of the jaws and, finally, becoming numb and feigning

When its underground home is endangered, the little American burrowing owl imitates very accurately the threat sound of a rattlesnake to ward off an intruder.

death are the final resorts.

Sounds employed for defensive purposes are not restricted to vertebrates. For instance, when disturbed, wasps of the genus *Synoeca* start the whole nest swinging, so producing a dull, drumming noise. Then they will suddenly swarm out of the nest and begin drumming more loudly outside. They produce this sound by means of synchronized wing-beats.

Normally, the yellow-bellied toad (Bombina variegata) is well camouflaged (below left). However, if frightened, it assumes this extraordinary position, so revealing the warning colours on its belly (below).

BURROWS, ARMOUR AND SPINES

The abdomen of the land hermit-crab (Coenobita) is not armoured. For protection it carries around a snail shell.

Buried in the sand, this fish relies for its safety on the poor eyesight of its enemies.

There is hardly any method of defence devised by man which has not been used somewhere in the animal kingdom. Certain animals are better endowed by nature than others. Camouflage, warning colours and noises, deceptive patterns and mimicry, all entail a measure of bluff. Escape, in order to be successful, depends on speed and stamina. Concealment is far more simple, particularly when an animal can retreat into a burrow.

There are countless examples of animals that find safety in shelter, notably among the rodents and reptiles. Many of the social insects, too, such as ants and bees, live in such retreats, sealing and guarding the entrances. There are also certain tube-dwelling worms living on the sea floor which can withdraw into their burrows with lightning speed when threatened.

Other species do not even need to go to such trouble, for they carry protective armour around with them; in most instances this consists of materials produced by their own body. Obvious examples are tortoises, armadillos and crustaceans. Some, however, find protection elsewhere. Hermit crabs use the empty shells of molluscs, and certain species go a step further by carrying around sea-anemones, the sting-cells of which keep intruders at bay.

Another effective method of defence is to carry spines. Those of hedgehogs and porcupines are virtually impossible to penetrate and can inflict severe injury, while those of certain sea-urchins and tropical dragonfishes are lethally poisonous and therefore provide excellent defence against inquisitive predators.

Barricade, bee-style: African bees of the genus Xylocopa *close the entrance of their tree-nest with the hard cuticle of* *their abdominal segments. Unlike honey-bees, these bees do not live in colonies, but often build their nests in groups.*

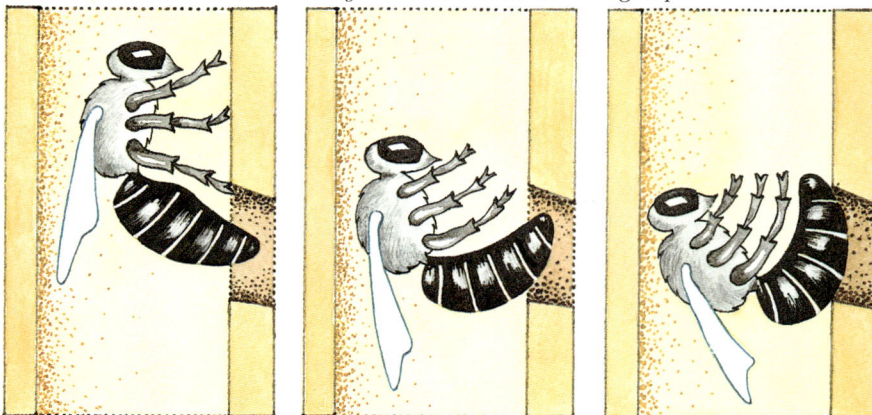

Box turtles have a transverse hinge in the middle of the belly armour, with which they can close their "box" up to the edge of the dorsal armour. They have thus developed to perfection a purely passive form of defence.

The Australian echidna is one of the evolutionarily primitive egg-laying mammals. In addition to being able to curl up, it possesses a poison spur on the feet, which protects the underside. Echidnas are not related to hedgehogs. Their spiny cladding is a case of parallel evolution.

South American armadillos (Dasypodidae) have a flexible bony armour. When rolled up, they are well protected against the jaguar.

The one-horn Indian rhinoceros (Rhinoceros unicornis) has primarily developed its powerful armoured plates as protection against the attacks of tigers. Nowadays, both the hunter and the victim are endangered.

Barricade, ant-style. Among the ant species Colobopsis truncatus are special door-keeper ants with a bung-like head.

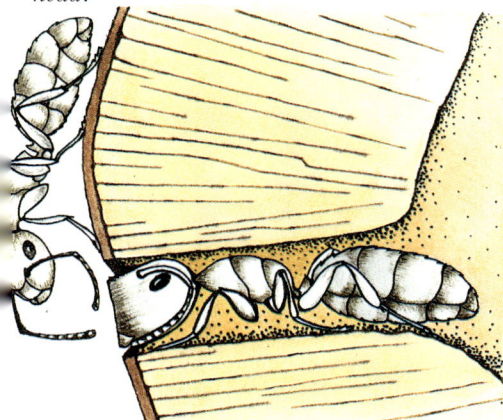

MAKING A SPEEDY ESCAPE

A jackal can put a whole flamingo colony to flight. This scene took place on the lake in the Ngorongoro crater. Jackals mainly attack flamingos at night when they wade out to rest in shallow water. Though relatively large, these birds, with their beak designed to filter food, are completely defenceless.

Gnus graze on the open plains. They are always vigilant and ready to make off at the approach of a predator. The herds stir up huge clouds of dust as they go. They have some of the outward characteristics of both the horse and the buffalo, yet can outstrip either for speed, reaching up to 80 km per hour.

In a zoo giraffes look slow and ungainly, but this is deceptive, because of their cramped surroundings. In the wild, a herd of galloping giraffes is a breathtaking sight. The long legs move rhythmically and the equally long neck swings back and forth. Giraffes can reach a speed of 50 km per hour.

Savanna zebras are the last wild horses to survive in large numbers. They are highly specialized runners, touching the ground only with the single toe that forms the hoof. Apart from man, their principal enemy is the lion, which usually overpowers them as they drink at night.

A healthy hare can escape quite easily from any of its enemies. With a top speed of 60 km per hour, it is one of the fastest animals. In addition, the hare has the astonishing ability, when running at full speed, of suddenly swerving and dashing off in another direction. This manoeuvre is usually too much even for the most agile and determined pursuer.

The astonishing ability of the hare to escape its hunters is due to its long, powerful hind legs. When running or jumping, it takes off with both hind legs at the same time, then brings both front legs back behind them.

Few animals have to contend with so many enemies. Hares have only survived thanks to their speed and agility, and the fact that they breed in enormous numbers.

FATAL BEAUTY

Even a foul-smelling or poisonous animal needs to advertise the fact that it is inedible, otherwise it will be eaten. Warning colours serve to do this. The more glaring they are, the easier it is for them to be recognized and respected by potential enemies. *Above*: the nudibranch snail Chromodoris luteorosea.

Nudibranch snails are easily recognizable by the ring of feathery gills encircling the anus, which collect oxygen. *Above*: the blue-and-yellow markings of Hypselodoris valenciennesi.

Left: the mollusc Flabellina affinis dazzles like an illuminated sign. The tentacles grasp captured sea-anemones with their poisonous sting-cells.

Far left: it is not clear whether the yellow hairs of the caterpillar Dasychira pudibunda actually serve as warning coloration. Certainly the long bristles are sufficient to irritate any animal unwise enough to attempt a meal.

Left: the black-and-yellow pattern of the fire salamander gives warning of its poisonous skin secretions.

Far left: only the male of the spider Eresus niger are brightly coloured, the females being inconspicuous. This is not warning coloration – the spider is not poisonous – but mimics a ladybird which tastes unpleasant and is therefore not sought as prey.

Left: the poison of the frog Dendrobates typographicus is so powerful that it can even kill large animals.

Far left: the striped bug Graphosoma lineatum does not hide from birds but relies for protection on its typical bug smell.

Left: the zebra moray-eel (Echidna zebra) has a poisonous bite. In a tropical coral-reef the striped pattern serves as camouflage rather than to frighten enemies.

Far left: the behaviour of this large locust corresponds with its warning dress which is extremely striking in the semi-deserts of southern Africa. When molested, it does not try to flee but squirts a stream of evil-smelling liquid at its opponent.

ARTISTS IN SURVIVAL

Menacingly, the African praying mantis rears its arms against the emerald cuckoo which is threatening it. At the same time, by spreading its front wings it brings the eye-pattern into play. Naturally the insect has no chance if the bird really decides to eat it; but the bluff is very effective, increasing the insect's chance of survival.

Right: when attacked, puffer-fishes have the ability to inflate grossly by swallowing air or water. A few species also possess pointed spines embedded in the skin. Pufferfishes likewise swell themselves up when they are trapped at low tide in shallow coastal pools. The stored air can partially be transported to the gill chamber where it is used in respiration.

The process of inflation works in the following way. The air that is breathed in is stored in an extensible fold of the stomach; circular muscles function as valves, preventing the air and gas from entering the stomach and oesophagus. The ability of the pufferfish to swell into a ball-shape is due to the fact that it has neither ribs nor a pelvic girdle.

ATTACK IS THE BEST DEFENCE

It is well known that many snakes are venomous, but not that certain non-venomous species such as the viperine snake (below) defend themselves by means of glands that secrete a foul-smelling liquid.

Although hunted animals frequently engage in all manner of tricks and ruses to deceive and confuse their enemies, many are quite capable of defending themselves directly. The predator is not always so superior in strength and skill that it can avoid being wounded by the weapons employed by the prey. The formidable horns of an African buffalo are capable of sealing the fate of a lion. A single blow from the steel-hard hooves of the ungainly giraffe can break the back of a carnivore. Huntsmen are well aware, too, that when cornered by hounds, chamois and deer will use their horns and antlers to defend themselves very effectively. In such a tight situation, chamois attempt to back up against a protective surface, such as a rock wall, while deer often face their pursuers on the edge of a lake, retreating slowly, if need be, into the water. Provided the teeth and claws of the predator can be avoided, an animal capable of butting or kicking stands a fair chance of saving its skin.

There are, of course, small, weak animals for whom defence by sheer physical force would be absolutely senseless and futile. Yet they, too, have methods of resistance. Some, for example, have evolved chemical weapons which can prove very unpleasant to an attacker, such as venomous stings in the case of bees and wasps.

Left: this African butterfly caterpillar seems to be oozing blood. When molested, many animals produce protective blood-like fluids. This is particularly astonishing in the case of insects, which have colourless blood.

Below: the leopard has brought down a male baboon. However, before it can carry its prey away, it is attacked by several of the victim's fellow baboons. Finally, it lets the prey go and moves off, followed by the shrieking and barking baboons.

However, these are not purely defensive weapons. Many wasp species use their venom primarily for attack. Some ants, probably evolved from wasp-like ancestors, also have a sting; others spray formic acid as their form of self-defence.

The bombardier beetle stores two chemical substances at the rear end of its abdomen which are relatively harmless when kept separate. If obliged to defend itself, this beetle brings the two substances together into a special chamber at the tip of its abdomen. An uncommonly strong chemical reaction produces an intensely hot liquid which is sprayed at the attacker.

Left: when a skunk is disturbed, it stops, raises the rear part of its body and emits a nauseous fluid, secreted by its anal glands.

Following pages. Left: apart from birds, enemies of red wood ants are mainly other ants and also beetles, ant-lions and spiders. The formic acid, squirted to a considerable distance, is particularly effective.

Right: the climax of the male gorilla's display to impress other animals is to strike the chest with the flat of the hand (not the fist) to produce a loud drumming sound. Among young gorillas, as here, the same behaviour denotes an invitation to play.

67

CAMOUFLAGE AND OTHER MARVELS

The human capacity for marvel is a rewarding and potentially creative attribute. Yet the majority of us tend to go about our daily routine unaware of the wonders around us. The truly great discoverers and innovators are those who marvel at something that to everyone else appears common-

survival animals are able to adapt themselves to a vast array of natural phenomena. Many contrivances in the animal world are so remarkable and intricate that they can truly be described as living marvels. The earth, in fact, contains as many marvels as it has animal species. Beyond the

Left: the Texas horned lizard (Phrynosoma cornutum) illustrates very clearly two principles of camouflage: adaptation of colour and body pattern to the structure of the surroundings. Under such circumstances, apparently conspicuous patterns have the effect of completely dissolving the outline of an animal's body. The human eye can no longer perceive the animal as a whole and so does not notice it.

Giant eyes for seeing in the twilight and flat suction toes for gripping smooth surfaces. Like this tropical tree-frog (Hyla calcarata), every animal possesses specific features of astonishing technical perfection. None is completely helpless. Each possesses at least one organ, one structure or one capability that gives it a particular advantage over other animals.

place. For many people a marvel has to be defined as an occurrence that defies the laws of nature. Yet a true marvel depends on the ability to exploit and master natural laws.

Animals are unsurpassed in the art of using nature's laws and turning them to their own advantage. In so doing, they have no need of instruction. Not for them the textbooks which are for us indispensable. In their struggle for

confines of the delicate biosphere that shrouds our earth, it is reasonable to assume that there is nothing comparable. Yet we treat this fragile and vulnerable living world as if it were something that cannot be demolished. We destroy and condemn to extinction the animals that breathe its atmosphere and with every animal species we exterminate, we take away something of ourselves.

THE FABULOUS CHAMELEONS

Right: a characteristic of chameleons is that they can move each eye quite independently of the other.

Below left: when a common chameleon encounters a snake, it inflates its throat sac, develops colour spots on the skin and makes growling noises.

Below, centre: a male four-horned chameleon (Chamaeleo quadricornis) in its courtship coloration. This West African species lives at altitudes of 1700–2000m.

Below right: this mountain chameleon (Chamaeleo montium) takes on bright colours when angry. In normal situations it is well camouflaged, and is best observed at night by torchlight.

In the realms of mythology the chameleon has an unjustifiably bad reputation, being pilloried in certain African tribal cultures as the bearer of evil – only one of the countless accusations that superstition levels against these small and completely harmless lizards. The very name of the creature, derived from the Greek meaning 'ground lion' implies a tacit misun-

that they can be turned quite independently of one another, affording all-round vision but focusing to provide precise binocular perception when prey is within reach.

The second unusual characteristic is the chameleon's proverbial ability to change colour. It used to be thought that a chameleon could adapt its coloration to any kind of background. Nowadays this is known not to be the case. It appears that camouflage is not really the main purpose of the colour change and that it serves mainly to regulate the body temperature. During the morning, it is the chameleon's first preoccupation to warm up. The body, which is

Facing page: close-up of a small Wiedersheim's chameleon (Chamaeleo wiedersheimi). The eyes of chameleons are remarkable in several ways. The circular lid leaves only a small opening for the pupil. The curvature of the lens enables the animal to estimate its distance from the prey.

derstanding of the nature of these bizarre but extremely interesting animals. The prejudice and mistrust may arise from at least three unusual attributes of these little, arboreal, insect-eating reptiles. In the first place, their large eyes are enclosed by circular lids to leave only a small, central aperture, and are mounted on twin turrets so

pale during the night, turns a darker colour in order to absorb the sun's rays as fully and effectively as possible. If the body becomes too hot, it becomes pale again, so as to reflect the excess rays. The ability to change colour also seems to be associated with mood, serving as a signal to other members of its species.

HUNTING WITH
THE TONGUE

Below: a territorial fight be-tween two mountain chameleons will continue until the weaker turns a dark colour as a sign of submission. The victor displays its full colour and raises its dorsal crest.

The third unusual feature of the chameleon is the ability to shoot out its tongue to catch insects, sometimes for a distance greater than its body length. The sticky tip of the flexible tongue transfixes the prey, which is drawn back into the

mouth. This ingenious device is based on the action of two groups of muscles. At the moment the tongue shoots forward, its powerful circular muscles contract and its longitudinal muscles relax. The result of this muscular interplay is for the entire tongue to unfurl and become elongated at lightning speed, as if being squeezed out like the contents of a tube. When the fleshy tip of the tongue has gripped and immobilized the prey, the longitudinal muscles come into play and draw the tongue back into the jaws, while the circular muscles relax. The champion in this, Fischer's chameleon, (*Chamaeleo fischeri*) can project its tongue a distance one and a half times the length of its body and tail.

The entire hunting process occurs too swiftly for the human eye to discern and can only be captured by the high-speed camera. But the mere appearance of the reptile, in repose, especially the male with his crest or frontal horn, conveys a 'prehistoric' aura of superstitious dread. Yet, the sinister reputation of the chameleon is based on nothing more than a series of highly developed adaptations to the way of life of an arboreal insect-eater.

Above: the African chameleon (Chamaeleo africanus) *can shoot out its tongue for a distance greater than the length of its own body, tail included.*

Left: in this picture sequence a chameleon locates its prey, calculates the distance and strikes, the circular muscles forcing out the tongue.

The tip of the tongue grasps the victim and . . .

. . . pulls it back into the mouth. The tongue is withdrawn as the longitudinal muscles contract and the circular muscles simultaneously relax.

VERSATILE EYES

Compound eyes are remarkably efficient for small creatures such as insects. Larger animals see better with lens eyes. If we had compound eyes, with the same degree of resolution as is provided by single-lens eyes, they would measure at least one metre across.

Top: like all its fellow species, the leopard gecko has excellent eyes which react primarily to the rapid movements of its insect prey.

Centre, above: the old belief that snakes hypnotize their victim is scientifically untenable. The fixed stare is due to the absence of eyelids.

Centre, below: birds have the most highly developed of all animal eyes. For instance, this gannet sees the world and its prey much more sharply than we do.

Bottom: frogs' eyes are remarkably flexible. In the tadpole stage they are adapted for seeing under water. The adult frog has to develop eyes which can also see in the open air.

The primitive eye probably consisted of a simple pit or groove containing a light-sensitive cell. Such a structure could possibly distinguish various degrees of light intensity and identify, very roughly, the direction of the incident light, but would not have been capable of perceiving colour or form. Lower orders of animals therefore have comparatively poor and limited powers of vision.

Highly developed eyes, by contrast, not only have an enormous number of sensitive cells, but also have complicated light-refracting structures, such as lenses and sometimes even mirrors. The visual system is not self-sufficient, however, for the visual information has to be interpreted very rapidly by a part of the brain. An eye, therefore, is much more than a camera, which cannot draw any conclusions from the pictures it takes. The larger the number of sensitive cells, the greater the capacity of the eye to see improved images. Many animals, including humans, have a combination of cells in the retina, known as rods and cones, and it is the latter which make it possible for the eye to distinguish colours.

Facing page: rear view of the small dragonfly (Lestes vireus) shows clearly the two eye systems of insects: at the sides of the head are the two giant compound eyes, in the middle of the head the three small ocelli.

76

Although a land crab is only about 3–4cm across, its stalked eyes have about 12,000 facets, almost as many as a 30cm-long lobster. Crabs can distinguish colours and the polarization of the sky. The eyes follow moving objects, so that their image on the retina remains stationary.

Underwater vision has specific problems, and the structure of a fish eye has to take account of this. Thus, in very clear water, the distance of vision is barely 40m. With increasing depth the light becomes scarce and some colours disappear completely. Furthermore, refraction is quite different under water. So the eyes of fishes are highly sensitive to light, with a strongly refractive spherical lens.

No snail has better developed eyes than the apple snail. They are borne at the tips of its tentacles, measure 0.3mm across, have a spherical lens and can, to a certain extent, discern forms. Their sensitivity to light is astonishingly high.

Cuttlefishes are cephalopod molluscs, related to octopuses and squids. All cephalopods have large, well-developed eyes. The eye of a deep-sea squid is as big as a soup plate and is, in fact, the largest eye in the whole animal kingdom.

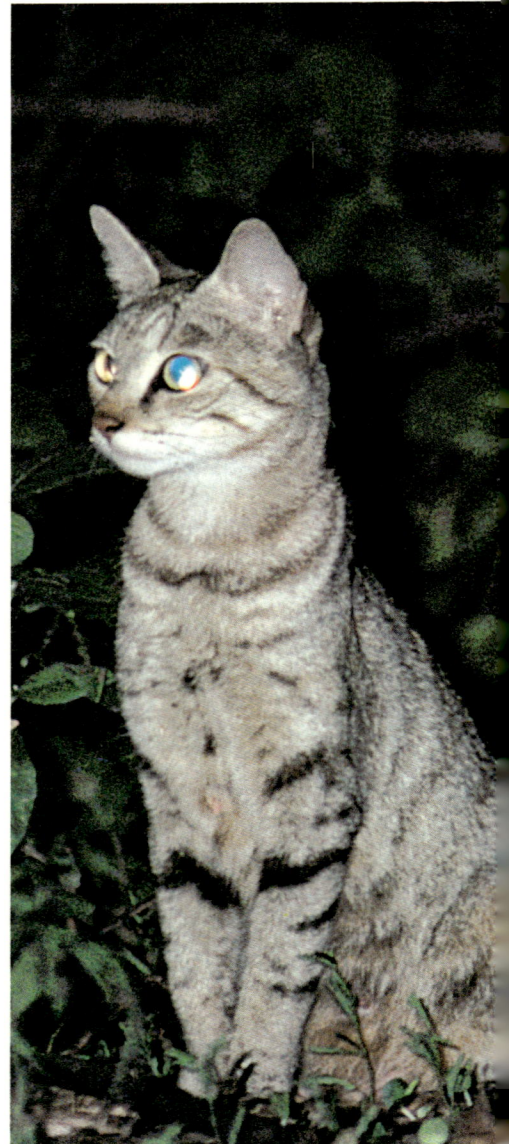

Many nocturnal animals have eyes that shine by torchlight, due to a special reflecting layer at the back of the eye, the tapetum lucidum. This reflection helps incident light rays to pass a second time through the sense cells, so making better use of what little light exists at night.

Eye of a horse-fly. The compound eyes of insects consist of hundreds to thousands of individual eyes, the ommatidia, each perceiving only one point of light. The light points of all the ommatidia together produce a kind of picture. Such eyes give insects astonishing powers of vision: they see colours, polarized light and flickering light better than humans.

The red-eyed tree frog of South America spends the night among branches in the rain forest. Its large eyes are adapted to make best possible use of the sparse nocturnal light. Frogs are the favourite prey of many predators. Their eyes are, therefore, so positioned on the head that they can watch in all directions. Many typical prey animals have such wide-angle eyes.

The small chikra hawk (Accipiter badius) also has bright red eyes. However, their structure is completely different from those of the red-eyed tree frog. Because they live by hunting, birds of prey have eyes which could be compared with telephoto lenses: they perceive only a small part of their surroundings, but with maximal sharpness and resolution.

In contrast to insects, arachnids such as this jumping spider have no compound eyes but a row of eyes with lenses. They cannot compete with insects in the resolution of rapid movements or flickering. However, spiders' eyes are superior in twilight and at night. Many spiders thus go hunting for insects when the prey can no longer see.

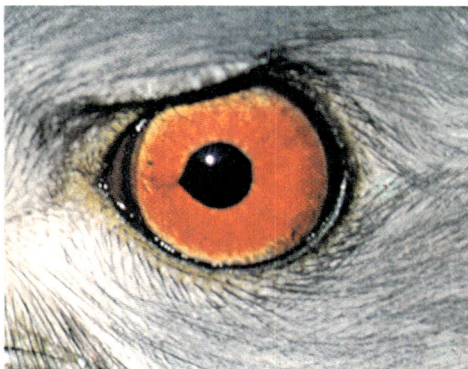

Colours play an important role in the courtship of bower-birds. These birds assemble collections of coloured objects in their bowers, in order to impress the females. The bright violet coloration of the iris has little to do with vision. It is a signal for the females.

In the red-footed booby the area around the eyes is, by chance, similar in colour to that of the bower-bird's iris. This coloration also serves mainly as a signal. Eyes are to see and be seen: in the animal kingdom such double functions are the rule rather than an exception.

NATURAL CAMOUFLAGE

*Above right: The larva of the fly bug (*Reduvius personatus*) is a perfect example of adaptation to the background. It camouflages itself with dust and soil particles, scraping the material together with its legs and throwing it over its back.*

The South American praying mantis Choeradodis *(left) and the Asiatic leaf-insect* Phyllium *(right) are model examples of the camouflage method known as mimicry. Mimetic animals imitate objects which are unimportant for predators, in this case green leaves.*

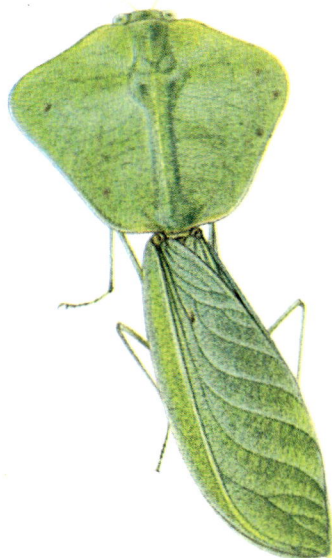

Escaping from a hunter is exhausting. For many prey animals it is simpler to avoid being seen. They do this by making themselves virtually invisible.

One common type of camouflage is adaptation of colour and pattern to the surroundings. Examples are white polar bears, arctic hares and foxes, ermines and ptarmigan in the snow; green

break the outline of an animal's body. This works particularly well against an irregular background, such as the bark of a tree or the gravel of a beach, or in dappled surroundings of light and shade. But in the open, or in uniformly coloured surroundings, such patterns, as in giraffes or zebras on the savannah, are often conspicuous.

Even more fascinating is the method whereby certain animals imitate things which are of no interest to the predator. This is a well-nigh perfect example of mimicry. Among the most astounding examples are the stick-insect *Carausius* and the leaf-insect *Phyllium*, both of which merge perfectly with twigs and foliage. In East Africa there is a "plant" with yellow veins on the petals. But an attempt to pick it or land on it may

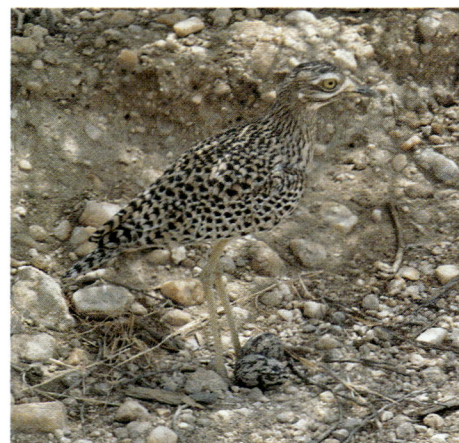

*Far right: the almost perfect camouflage, by colour and pattern, of the stone-curlew (*Burhinus*) is only surpassed by its egg clutch which lies, almost invisible, directly on the ground.*

caterpillars, beetles and bugs in the treetops; and sand-coloured lizards, snakes and mammals in the desert. Just as effective is the possession of a suitable pattern of coloured spots and streaks which

result in the flower flying away. Here the mimic is a species of cicada which may be wrapped so convincingly around an upright stem that even botanists are deceived.

The South American moth Thysania agrippina, *with a wingspan of 30cm, is the largest of all moths. In spite of its size, this moth is difficult to discern, because the pattern of its outspread wings, as it rests on a tree trunk, perfectly imitates the bark. To the left of the giant moth is the bug* Dryoctenes, *which in its way also imitates the bark structure.*

The frog Megophrys *lives in leaf litter of the rain-forests of South-east Asia. With its improbable eyes and nasal tips, which are very sensitive, it looks remarkably like a dead leaf.*

ODOURLESS,
MOTIONLESS,
INVISIBLE

Above: the newborn Thomson's gazelle is not only almost invisible but also gives out hardly any scent.

Right: the larva of the tortoise beetle (Passida) carries its old larval skin, smeared with faeces, on its back. This seems to serve not only as camouflage, but also to frighten away ants.

Centre, right: when little tern chicks duck down behind stones, they blend in with their surroundings.

An animal does not have to make itself invisible in order to go unnoticed by predators. Roe kids and antelope calves, for example, camouflage themselves not only by blending with their surroundings, but also by being almost odourless.

Whatever the type of camou-flage, however, it would not work unless it were accompanied by appropriate behaviour. Movement would immediately be fatal. Thus, young hoofed animals instinctively remain motionless. Chameleons, spiders and insects are also experts in immobility. But

Above: When it lifts its head, this gecko cannot be distinguished from the branch on which it is lying. The success of camouflage depends as much on an animal's behaviour as on its coloration.

different environments. Flatfishes are especially adept at doing this. By visual reconnaissance, they take on the coloration of the sea floor, while visual nerve impulses direct the pigment cells to change so that the pattern of the sea bed is also simulated.

Some animals camouflage themselves by using materials from their surroundings. The South American fly bug powders itself

Above: in the foliage of a treetop the sun shining on the leopard's spots provides perfect camouflage.

Left: a case of mimicry: the motionless praying mantis (Empusa) resembles a twig.

conversely, it is no good merely staying still if the surroundings are exposed: a bark-coloured moth would be very obvious on a green leaf so a suitable background has to be found. Some species are particularly fortunate in being able to adapt the body colour to

with sand, and various crab species cover their shells with algae, sponges and pieces of coral. A green lacewing larva, which feeds on aphids, envelopes itself in the fluffy, waxy discharges produced by its prey to avoid being seen by the ants which protect the aphids.

Above: inside the flower, perfectly colour-matched, lurks a crab spider (Thomisus) waiting for unsuspecting nectar-feeders such as bees, which it kills and sucks dry.

Far left: this iguana blends remarkably with the surrounding terrain.

FEET THAT GRIP

Right: colonies of the African rock hyrax (Heterohyrax brucei) usually live in cliff crevices and deserted termitaria. However, this small insectivorous and plant-eating animal also climbs well on tree trunks. The related tree hyrax (Dendrohyrax) lives almost exclusively in trees.

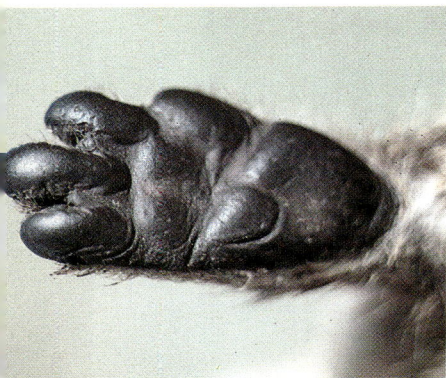

Above: the secret of the hyrax's climbing prowess is a foot sole adapted to any surface unevenness.

Far right: the foot of the rare palm gecko of New Guinea from below: there are microscopic hooklets on the lamellae.

Facing page: the larva of the emperor moth clings to a plant stem. Clearly visible are the numerous small processes of the caterpillar's foot, which greatly enlarge the foot surface and thus enable it to cling by adhesion.

Reflecting their life styles, burrowing animals have feet equipped for digging, aquatic animals possess paddle-like or webbed feet for swimming, sloths hang upside-down from branches with their hook-like feet, carnivores are armed with sharp, curving claws, and fast-moving herbivores that live on plains have very hard hooves with only a small surface area. Among the most specialized forms are the suctional feet of many climbing animals. In these examples the method of attachment to the surface varies; a fly descending head downwards on a window pane is held firm by adhesion. There are small projections with thousands of tiny hairs on the feet which grip every slight unevenness. The toes of geckoes each have a pad of ridge-like scales; each scale bears hundreds of hairs with multiple hook-like tips that grasp the tiniest irregularities.

MATING
BEHAVIOUR

The survival of a species depends on its ability to produce offspring in sufficient numbers, generation after generation. Because of this biological need, mating is the most important event in the life of an animal. It releases very powerful sexual drives and the most astonishing patterns of behaviour, both young, establish sophisticated food stores or seek hosts upon which their own larvae can feed and thrive as parasites. There are fishes which migrate thousands of kilometres to reach their accustomed breeding grounds and others which keep their eggs and fry protected in the parent's mouth.

Left: gorillas have a very intimate family life. Even adult males enjoy the antics of the young and play with them. Care of the baby, however, is not inborn, instinctive behaviour for the gorilla mother. She learns it as a youngster by watching the family group. In the zoo, young gorilla mothers, who have not had this experience, do not know how to deal with their offspring, letting it starve or even killing it. The baby therefore has to be removed and reared artificially.

Right: the mating activity of snakes often continues for a long time, with a tender and almost dance-like courtship.

in terms of finding partners and caring for the newborn young. As a general rule, the smaller the animal, the more offspring it produces, for many, if not most will be lost. At the end of their larval stage, some insects cease to feed and are only concerned to mate and to lay eggs. Many build architectural masterpieces for their Female spiders may be bloodthirsty murderers of their mates and at the same time behave like model mothers. During the greater part of the year many birds are primarily concerned with caring for their brood, and there are certain mammals which have a true family life that in some cases can be genuinely affectionate.

COURTSHIP
DRESS AND
BRIDAL GIFTS

Above: many male birds present their mates with wedding presents. In the case of the shag, a small species of cormorant, the gift comes in the form of nest material. In other birds it is a beakful of food. Both symbolize the readiness of the male to concern himself with the wellbeing of the future family.

Facing page: in India, for thousands of years, the peacock has been the symbol of Krishna. In pre-Christian times Persia had its peacock throne; and in Greek mythology the peacock was dedicated to Hera, queen of the gods. The male peacock possesses the most beautiful plumage of all birds. The erect, widely spread tail coverts play a part in mating ritual, being designed to impress.

It is obvious from the mating antics of dogs or the aggressive courtship of domestic fowl, to take only two commonplace examples, that sexual union can be a highly complicated process. In most animal species, elaborate preparations and often very sophisticated behaviour patterns are necessary before the actual stage of copulation is reached.

Among many animals, prominent colours and patterns, sometimes very different from those in evidence at other times of year, assume a highly important role. A male king penguin has orange-yellow patches on the sides of the neck. Experiment has shown that if these are painted over in black or white, the bird no longer has any chance of finding a mate. During courtship display certain fishes and lizards change colour entirely and become very attractive to prospective mates. Before the nesting season many male birds exchange their dull eclipse plumage for a beautiful breeding plumage which is very prominently displayed during courtship. In addition, the bill and feet of the naked skin or the head or neck, normally very inconspicuous, may become brightly coloured. In certain species, such as the peacock, the point of attraction is the erectile head crest or the conspicuous fan-like tail.

In other species the external characteristics hardly change, and the males resort to a variety of stratagems. Many male birds present the females with a gift such as food; or the male will collect twigs, straw and similar materials as a symbolic prelude to the construction of the nest.

The danceflies of the family Empididae have evolved a most remarkable range of courtship offerings. Some species in this family are predatory. At mating time the male brings a captured insect as a gift. Other dance-flies enclose the wedding present in a cocoon of silk before handing it over. Depending on the species concerned, the male will either suck out the contents of the silken package, or simply present the empty silken balloon. Whatever the form in which the gift is presented, the female proceeds to consume it in the course of copulation.

Right: in its breeding plumage, the puffin is a brilliantly coloured bird. In autumn it sheds the attractive, horny bill sheath, leaving a much smaller, brownish beak. The triangular pattern around the eye, consisting of dark horny plates, also disappears. The bright red feet become an inconspicuous greybrown colour.

Above: at mating time the plumage of the male Lapland bunting contrasts very strikingly with that of the female.

Below, centre: male and female divers are often hardly recognizable when the eclipse plumage gives way to the breeding plumage. Left: a pair of great northern divers. Right: a pair of black-throated divers.

1. Curled tail feather of the drake mallard in breeding and eclipse plumage.
2. Head feathers of the great crested grebe. In males and females, rust-red feathers form a large erectile crest.

In the ruff the male has a bright collar for its courtship display.

Only in its breeding plumage does the black tern take on a darkish hue.

The back of the dunlin becomes red-brown, the belly black.

The white wagtail becomes smarter before the breeding season.

3. Head feathers of the drake mallard. The iridescent blue-green of the breeding plumage becomes brownish with spots in the eclipse plumage, as in the female.

4. Head feathers of a brambling . . .
5. . . . . and of the black-headed gull in breeding and eclipse plumage.

Above: to attract the female, the male frigate bird inflates his throat sac to a bright red balloon, the size of a child's head.

GARDENS
AND BOWERS

Right: a golden bower-bird brings new decorative material for its maypole bower.

Below: a young satin bower-bird, still in juvenile plumage, is already practising the courtship dance.

Right: the regent bower-bird has enticed the female into the bower and immediately proceeds to mate with her (far right).

The Australian bower-birds are unique among the passerines for their ability to distinguish colours and use objects for decoration. The males of certain species build avenues and bowers, huts, maypole trees and courtship arenas. They clean the leaves and twigs away from an area on the forest floor. Some of them then erect a protective wall of thin twigs and moss. Among the maypole tree builders, the male places thousands of tiny twigs around the stem of a small tree, up to a height of several metres. The bower builders use interwoven grass to construct elaborate shelters, even daubing the inner walls with pigments, using bits of bark or leaves as tools. Others strew the ground with pebbles, glass, snail shells and other objects to entice the female. She builds a separate nest for laying eggs and rearing the chicks.

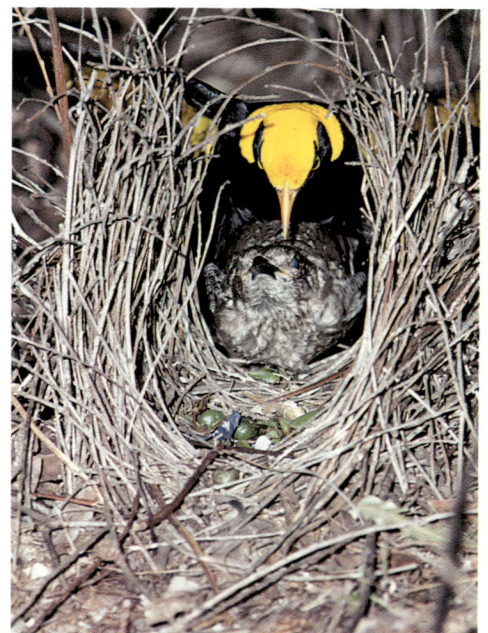

Facing page: the satin bower-bird likes to adorn the front of its bower with blue flowers and with butterfly and beetle wings. It also collects other blue objects, including, in this instance, clothes pegs from a nearby garden.

DANGEROUS
COURTSHIP

Above: at risk of his life, the male orb-web spider lands on the web of the female, causing it to swing. At the same time he signals to her with his legs.

Right: mantis courtship usually ends with the death of the male. Even in the course of mating the female starts to eat the front of her partner's body, while his abdomen completes the act.

which the front antenna-like limbs make special signals. The males of certain species are sophisticated enough to tether the female to the ground so that mating can proceed in peace. Others capture a prey animal, wrap it in silk and present it to the female who is then busy with the gift during mating. In yet other species the male grips the murderous poison claws of the female and keeps them open or closed during mating.

The danger, nevertheless, may not be confined to the approach stage or the actual mating activity; it may extend to the aftermath. A female spider's sexual drive may turn to aggression after mating. If the male does not disengage quickly enough, it will be eaten.

The courtship and mating of mantises are likewise potentially dangerous to the males. The approach has to be conducted with the utmost caution. The male may

In most spider species the females are considerably larger than the males, and they are so predatory that they attack practically any smaller animal that is nearby and kill it by a poisonous bite – even unwary males of their own species. Male spiders, however, have a few ingenious tricks to avoid death at mating time.

Male orb-web spiders vibrate the female's web to signify that it is not prey but a sexually mature male. In spiders without a web the male performs a courtship dance which may last for hours, during

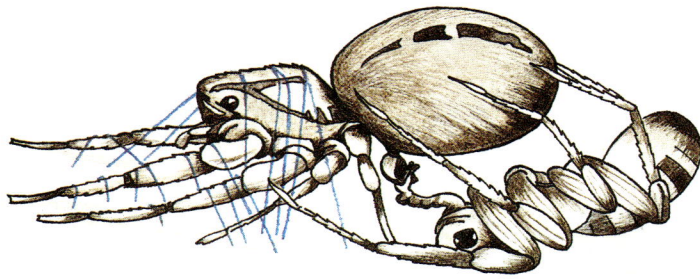

Left: the male of various spider species is extremely adept in pinning down the female, quickly mating before she releases herself.

take an hour or more to creep up, unnoticed, choosing his moment to clasp the female's body, preparatory to mating. If, while copulating, she sees him or is distracted, she will bite off his head. Even in this unhappy state, the rest of his body will continue to perform the sexual function until completed.

Above: a female giant orb-web spider from Florida has caught a large dragonfly. While she is occupied with feeding, the tiny male uses the opportunity for mating.

Below: male jumping spiders sometimes have to dance before the female for hours and make signals with their front limbs so that they are recognized as members of the same species, thus producing the necessary stimulus for mating.

DANCE, SONG
AND LIGHT
SIGNALS

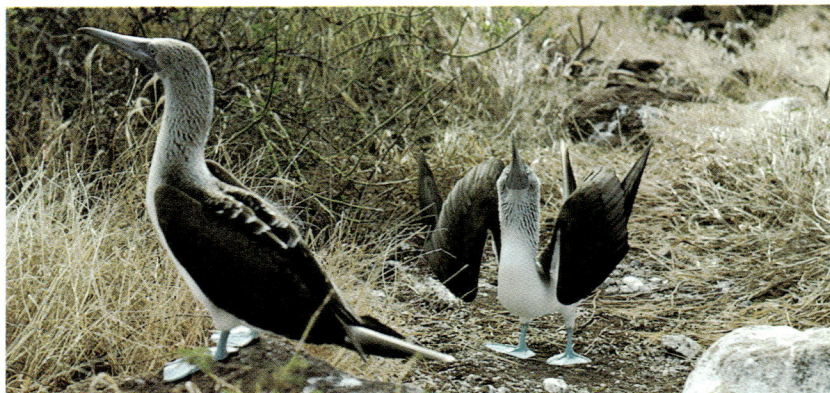

The start (above) of the long and complicated courtship of blue-footed boobies, relatives of the gannet. The male booby bends forwards, stretches his neck and bill upwards, and flutters his half-open wings. This procedure resembles the begging movements of the young, stimulating the female's maternal instinct and making her ready to mate. Then comes a mutual courtship dance with formal steps, beak scissoring and symbolic nesting gestures (centre pictures). In this repetitive ritual the movements of each partner always match each other exactly. It appears that mating (bottom) is not possible without this synchronized activity.

Animals have developed an astonishing array of methods for enticing a partner and establishing the right mood for mating, as has already been noted in the case of the peacock and bower-birds.

Mating dances occur in the courtship repertoire of spiders, fishes and particularly birds. In various aquatic and fowl species, male birds assemble at hereditary jousting places and perform warlike dances in front of the apparently indifferent females. Male and female divers, boobies and albatrosses often dance in pairs for days at a time, sometimes even causing themselves injury. By synchronizing their movements, they work themselves up to the necessary pitch of readiness to copulate. Such elaborate displays would not be necessary among species that do not form lasting pair bonds.

Song birds use special calls in order to keep potential rivals away from their breeding territory and at the same time to entice unattached females. The croaking of frogs, the chirping of grasshoppers, crickets and cicadas, the buzzing of mosquitoes and the sounds made by fishes all serve to attract partners for mating. Many deep-sea fishes have light organs that differ in detail from species to species, helping them to recognize individuals of their own kind. They probably also play a role in pair formation.

This is certainly the case in fireflies which flash light signals to one another; these sometimes have a rhythm specific to a particular species, this being designed to help the partners to identify each other and come together for mating in the darkness.

Male deer lay claim to their territory with a roar audible over several kilometres: one of the most impressive sounds in the animal kingdom.

Hummingbirds perform spiral courtship flights, repeatedly coming together in the air for several seconds.

MASS PRODUCTION AND BROOD PROTECTION

Right: a clutch of bug eggs. Insects arrange their eggs in a well-defined, species-specific pattern. The eggs must not be too conspicuous so as not to attract parasites and enemies that feed on them.

Far right: the eggs of the water stick-insect Ranatra *are laid below the surface in soft plant material. Each has two respiratory processes, waving free in the water, which supply the egg with oxygen.*

Below: sequence showing the mating of the midwife toad and the string of eggs being taken over by the male.

Start of mating, male gripping female.

Right: the eggs of empid flies are laid singly, each on its own stalk, because the hatching larvae are cannibalistic and in a tightly packed clutch would feed on one another.

Amphibians, like reptiles, reproduce by laying eggs, and in vast numbers. Only an infinitesimally small proportion of these will hatch and produce tadpoles. Thus, a single toad lays about 6000 eggs. On average, only one young toad survives to reach sexual maturity. The others are mostly eaten during the larval stages or succumb to other accidents.

The majority of frogs and toads use this strategy of mass-producing eggs. After the eggs are laid, the adults usually go off, providing no form of brood care, leaving the spawn to its fate. Yet there are exceptions to the rule. Several species have developed very astonishing methods of protection. The Brazilian leaf-frog builds a wall of mud in shallow water, extending above the surface. Here the tadpoles are protected against swimming predators. The male of the European midwife toad, a species that lives in holes close to

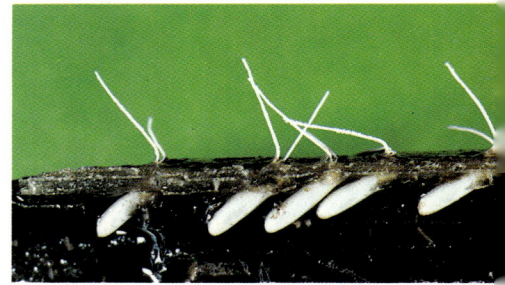

water, fertilizes the strings of eggs and then entwines them around its hind legs, carrying them about with him for weeks, keeping them moist until they are ready to hatch. Male tree-frogs guard the eggs laid on moist ground under stones by the females. After hatching, the tadpoles wriggle on to the back of the father, hanging on by means of a sticky skin secretion, and absorb-

Egg-laying and fertilization.

The male takes over the eggs.

ing oxygen through their skin. Some days later the father deposits them in the nearby water. Tadpoles of another frog species of the Seychelles Islands behave similarly but remain on the father's back until they are fully grown, feeding on what is left of the eggs. Among other leaf-frogs, the eggs slip into a depression on the female's back, where they are protected. When the larvae have developed legs, she places them in the rain-filled leaf funnels of bromeliad orchids. The female marsupial frog has a skin pouch on her back in which the eggs and larvae are well-protected, and the eggs of the grotesque Surinam toad develop in small pits in the female's dorsal skin, from which the fully formed young emerge.

An emperor moth (Eudia pavonia) *laying its spiral egg clutch. Moths do not practise brood protection and so have to lay large numbers of eggs to compensate for the depredations of parasites and predators. The eggs are usually laid directly on the caterpillar's food plant so that after hatching they can start to feed immediately.*

The pair separates.

Only the male tends the eggs.

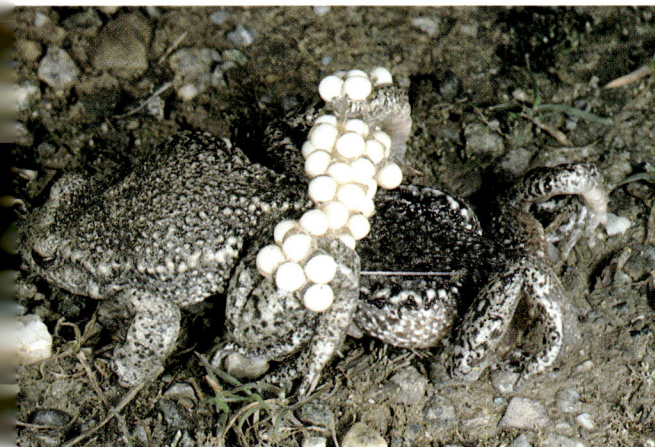

PICK-A-BACK

After it has spent 8–10 weeks in the pouch the young opossum climbs on to its mother's back and is carried around. It does not use its own tail to cling on to her tail.

The young of the flying lemur of South-east Asia, born virtually as embryos, attach themselves to a nipple shortly after birth. They remain there during the mother's gliding flights, until they are quite large.

Australian koalas which live in eucalyptus trees also practise pick-a-back. The young are born in an embryonic state, spend about six months in the maternal pouch and then get around by riding astride her back.

Cats, including lions, transport the young in their mouth to a new lair when they are threatened or disturbed. Evidently the young are not injured by the sharp teeth.

Many animal mothers protect their helpless babies by picking them up, generally by the scruff of the neck, and carrying them to safety. Other youngsters travel around with their mothers as a matter of course. Baby baboons, chimpanzees and macaques cling to the fur of their mothers' abdomens and later climb astride their backs. But monkeys are not alone in providing this sort of infant transport.

In the early 18th century the German artist Maria Sibylla Merian painted a South American opossum mother with a dozen or so youngsters on her back, their tiny tails wrapped around her extended tail. For more than a century, this illustration was widely copied, but the position of the mother's tail gradually came to be curved over her back. Scientifically, this has been shown to be incorrect. Many marsupials, such as this opossum, do carry their half-grown young on their back, but the young actually grip their fur with their claws.

Another small marsupial, the Australian koala, is even more renowned for its ability to carry young pick-a-back. The baby spends about six months in the mother's pouch, and then clings tightly to her back for a few more months as she clambers up and down the tree and along the branches, until it can get about on its own.

Birds cannot carry their young during flight. The weight would greatly restrict their ability to fly. The chicks of some aquatic birds,

Young bats hang on to the mother's belly and can even suck milk in flight. When they are bigger she leaves them behind in a hiding-place during her nocturnal hunting trips.

however, such as swans and grebes, sometimes clamber on to the back of the parents in order to rest, dry out and warm up. Bats, however, do fly around with their young while the latter are still quite small, on their nocturnal travels, until such time as the baby becomes too heavy; then it is left behind in the cave or tree until eventually it can itself take part in the hunting trips.

Various arthropod species carry their young on their backs. The best known are the scorpions. The babies are born alive or hatch immediately from the eggs after these are laid, and promptly clamber on to the mother's back. They could scarcely find better protection, because the female holds her tail, with its venomous sting, over them. If one of the young falls off, the female carefully places a claw over it so that it can climb back. Scorpion mothers will also adopt other babies that have strayed.

Some female spiders show dedication beyond the line of duty. While carrying their numerous offspring, they let their babies nibble away at them until they are completely devoured.

During their first weeks the babies of baboons and other monkeys hold on with hands and feet to the mother's chest fur. Later they ride on her back.

The chicks of the great northern diver and of many other waterfowl can swim – sometimes dive – shortly after hatching but they climb on the parent's back to rest, warm and dry among the feathers.

Young scorpions have scarcely hatched from their delicate egg shells before they climb on to the mother's back. She holds her venomously armed tail over them in a protective position.

SAFE AND WARM

Above: bank voles are born in a well-protected underground nest or sometimes even on the ground. If the mother is threatened while suckling her litter, she will make her escape with the babies still firmly attached to her nipples.

Right: many female domestic cats display superior instinct when seeking a safe refuge for their kittens. Only when the latter can stand on their feet will she take them out to a place where they can suckle in peace.

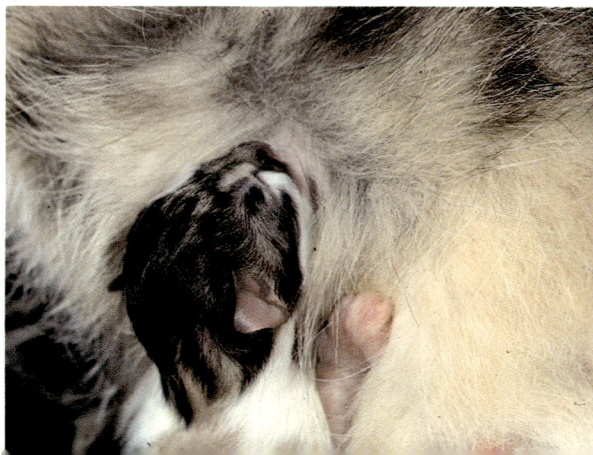

Left: like lions, the lemurs of Madagascar live in social groups, the young being reared not only by the mother but also by the community. They can suckle from any lactating female in the group.

Left: like most other young birds, stork chicks do not freeze in bad weather. They find ample protection from damp and cold under the wings and belly of the parents.

Above: polar bear cubs are born in a snow hole, usually at the beginning of December; they are only 30cm long, blind and helpless. During the whole winter the mother remains in the snowed-up lair, without feeding. Her body warmth maintains an adequate temperature and she lives on her thick layer of fat.

103

KANGAROOS

Facing page: as soon as a baby red kangaroo has left the pouch, the female produces a new one which immediately takes its place. To enable the older one, in spite of this, to continue suckling, the female henceforth produces different kinds and amounts of milk from different nipples. Even if a baby dies she can immediately, and without mating, replace the loss because the unborn one is already waiting in her uterus.

Right: a newborn red kangaroo is no larger than a worm. Despite this, it finds its way into the nourishing pouch, if necessary without the mother's help. If repelled, it begins the journey again.

Far right: whereas the young of higher mammals grow in the maternal womb, baby kangaroos go through their developmental stages in the pouch.

Human babies, when born, have already spent nine months maturing in the mother's womb. Nourished by the maternal placenta, they are equipped in some measure for life after birth. True, like the young of some other higher mammals, they are still dependent upon their parents, for a variable period, until they are mature, but they are far better developed at birth than newborn kangaroos. The baby kangaroo, in fact, is literally born twice. In the case of the red kangaroo, the first birth takes place about five weeks after fertilization. The baby, virtually an embryo measuring only about

animal finds its way on this long journey, which takes several minutes. It was once thought that the mother attempts to help by licking a path through the fur. However, recent observations have shown that the baby manages to reach its destination without any assistance. It spends seven and a half months growing and developing within the marsupium. Only then does the young one leave the maternal pouch: its second birth.

A female red kangaroo can rear up to three young at the same time. While the largest youngster is already moving about on its own, though still continuing to suckle,

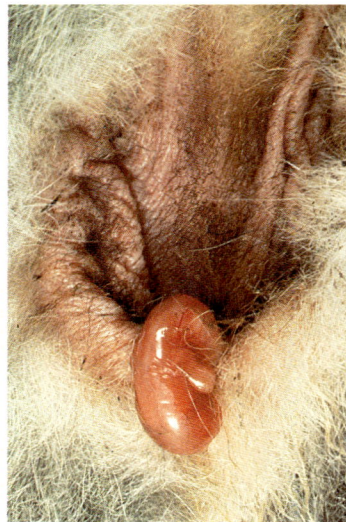

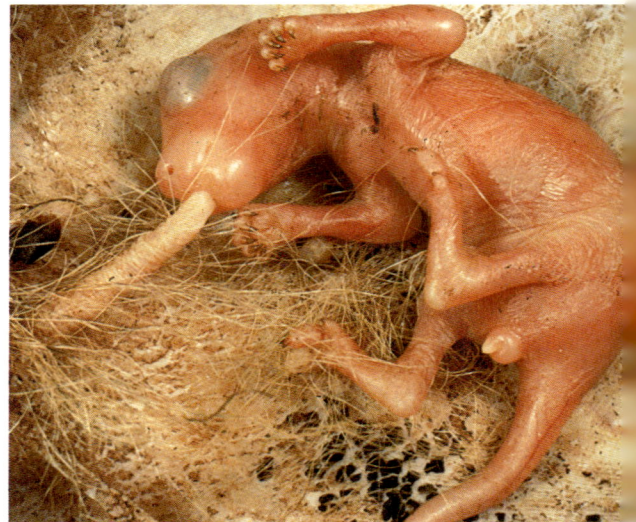

two centimetres, leaves the mother's womb and, of its own accord, climbs up the mother's body clinging to the abdominal fur, until it reaches the marsupium or pouch, where it immediately attaches itself to a nipple, not letting go of it for the next six months. It is still not known exactly how the tiny

its smaller sibling is growing in the marsupium and another embryo is waiting in the uterus, ready to move on when the pouch is free. The weight of the baby developing in the pouch may slow the mother down, especially if she is being hunted, in which case she will drop the baby out of the pouch.

UNCLES AND AUNTS

When attacked by predators, foals are protected by the whole community but are only tended by the mother.

Lion cubs command the affection of the whole pride. Once past the very early stages, it is difficult to decide which is their mother, because they can be suckled by any lactating female.

Most birds and mammals, and numerous other animal species, notably among the insects, care for their babies most assiduously. Understandably, this protective attitude is usually reserved for their own young. The offspring of neighbours are, as a rule, neither sheltered nor fed, indeed scarcely tolerated in the immediate vicinity, and may occasionally even be killed.

Yet there are several notable exceptions to this rule. Among elephants, for example, all the lar-

ger members of the herd take part in caring for the young. This attention is evident from the moment a baby is born. Adult females will assist at the birth and band together to protect the calves when danger threatens. If a baby elephant loses its mother it will be adopted and suckled by an aunt, who will rear it as her own.

A lioness in advanced pregnancy leaves the pride and gives birth to her young in privacy. After about six weeks, when the cub can walk, the lioness brings it

Left: each female guillemot lays only one egg, but in the breeding colonies the chicks soon gather in small groups which are tended by a single old bird, while the other adults are out fishing.

Below: in bad weather, emperor penguin chicks congregate in large, densely packed groups. Some of the adults form a protective wall while the others are often fishing over a hundred kilometres away.

to the pride. Here it will not only enjoy the protection of the other lionesses but may also suckle from any other lactating female. Even an adult male will tolerate the cub, allowing it to romp around, play with his twitching tail or take small scraps of meat from between his teeth.

In colonies of emperor penguins, a few adults will watch over multitudes of young birds while most of the parents are busy catching fish. Guillemot chicks are likewise tended by individual adults and another remarkable example is that of an East African wood hoopoe. These birds live in small groups and normally only one pair is fed and tended by all members of the group.

Young large-toothed hyraxes play not only with the young of their own species, but also with rock hyraxes; this must be unique among wild animals.

THE RESOURCEFUL HORNBILLS

Right: each young large-toothed hyrax has its own nipple, which it defends against the other young. When there are more than two offspring, the strongest take the two front nipples because these yield most milk.

Below: the blue-footed booby transports fish in its stomach to the nest. The strident begging of the chick stimulates it to open its bill, whereupon the baby thrusts its own bill into the parent's throat so that the food is regurgitated.

Below, centre: to build its brood cells the leaf-cutting bee (Megachile) uses pieces of leaf which it cuts out with its sharp mandibles. From the round leaf pieces it arranges rows of finger-like cells in wood or in hollow stems (far right). Nests of the leaf-cutting bee are also sometimes found in the soil.

Animals employ countless ingenious methods of protecting their young, both before and after birth. One of the strangest and most effective is that of the hornbills, widely distributed in Africa and in Southern Asia to New Guinea. Among the principal enemies of these birds are the slender, agile genets which feed to a large extent

way of combating such thieves: they wall in their nest. Once a pair of hornbills has selected a suitable hole in a tree, the male collects damp, preferably loamy soil or animal droppings. The female mixes this material with sticky saliva, plasters it around the entrance to the hole and presses it firm with her bill. Working from

on the eggs and chicks of birds. Larger bird species which breed in tree burrows with relatively wide entrances are defenceless against these nest robbers. But the hornbills, some the size of jackdaws, others as big as geese, have found a

the inside, she reduces the entrance to a small slit, through which she is continuously fed by the male. Within the safety of the wall, which becomes astonishingly hard, she incubates and hatches her eggs. When the chicks are a few

Left: the mother hippopotamus always keeps her baby close to her head or neck so that she can guard it and defend it against enemies such as crocodiles or lions.

Far left: after the food supply for the larval leaf-cutting bee has been brought in and an egg is laid in the cell, the latter is closed with a suitable piece of leaf. The larva can now grow, well supplied with all its needs.

Below, centre: the mason bee (Osmia bicolor) is related to the leaf-cutter bee but lays its brood in empty snail shells — usually three, four or even more cells in each shell.

weeks old, the mother breaks the walling and leaves the nest. Immediately both parents begin flying to and fro, collecting new building material to be used to repair the nest. This time the chicks themselves reduce the size of the entrance to the nest. This pattern of behaviour is presumably hereditary. The two parents can now bring food to the chicks in the nest, feeding them through the slit. When the chicks are fledged they also break out and fly off.

Above: a section of the mason bee's cell walls, made of masticated green leaves. The cells are provided with larval food and an egg. A layer of small stones is packed between the cells and finally the shell is camouflaged.

ANIMAL ALLIANCES

The evolution of animals is marked by gradual progression from simple organisms to ever more complex forms of life. Single-cell organisms come together to form colonies of cells and finally multicellular bodies. These again form associations of increasing complexity, culminating in the members of which play a similar role to the cells of the human body.

Some relationships between two animal species, which originally began as parasitism, whereby one lives at the expense of another, have changed into symbiosis, an active partnership either necessary or advantageous to both members

Left: a shoal of cardinal fishes sheltering among the spines of a sea-urchin (Diadema). These fishes use such shelter in order to feed on the planktonic organisms that drift past on the ocean currents. It is not known whether the sea-urchin profits from the presence of the fishes.

Right: pelicans and boobies nest together on the South American guano islands. This is not an instance of symbiosis. The birds live packed together owing to the scarcity of suitable breeding space. Mixed sea-bird colonies form a warning and defensive community.

highly developed colonies of termites, ants, wasps and bees. In these social communities the division of labour is so advanced, most of the members being engaged in food-gathering, building or defence, that only a few individuals – the sexual animals – are capable of reproducing. This is why these insect societies have been described as super-organisms, the in which the participants may be very different. In some cases the contrast is extreme. For instance, giant clams (*Tridacna*), up to two metres across, are nourished by the photosynthesis of tiny symbiotic algae. We ourselves have a symbiotic relationship with micro-organisms, namely the bacteria in the intestinal flora, without which digestion would not be possible.

THE SENTINELS

Right: the worker guard of a wood ant colony in defensive stance and, to its right, a normally positioned wood ant.

Above: marmots and hyraxes belong to completely different mammal groups, but both live in families and have look-outs which pipe shrill warnings to the members of the community whenever danger threatens.

Certain animal communities that are vulnerable to attack post sentinels to warn their companions of impending danger. The marmots are a case in point. The first sentinel lets out a shrill whistle to alert other grazing and playing marmots. The alarm signal is then immediately taken up by other sentries in the area, whereupon all the animals dive for shelter. A casual passer-by will hear the warning signal without seeing the guards. The marmots only retreat to their burrows for about half an hour; then they poke out their heads and, if the coast is clear, resume their activities.

The South African meerkats which live in earth burrows have a similar watch and warn behaviour pattern. Unlike the marmots, which are rodents, meerkats are small carnivores of the family Viverridae, related to mongooses, which feed on mice, lizards and insects. These predators are, however, comparatively small and so have to beware of larger ones. When on guard, meerkats stand on their hind legs in a characteristic pose, using the tail as an additional support. This gives them a good view of their immediate surroundings. Their worst enemies are vultures which, with their keen vision, spot them from a great height. It is interesting to note that meerkats have two different warning calls, one for a ground predator, the other for an enemy on the wing. The former is in part challenging and aggressive; the latter is simply an expression of fear.

Below: the topi (Damaliscus lunatus topi) *will often stand for hours on an elevated place, such as a termitarium. It is not clear whether it is on guard or whether it is simply demonstrating its possession of a territory.*

Left: various small animals of open grasslands survey the surroundings by standing on their hind legs, like the meerkat (Suricata), *a member of the family Viverridae.*

Far left: the African ground squirrel, similarly on the watch, is a rodent.

SAFETY IN NUMBERS

Facing page: individual cape buffaloes are often killed by lions, although they are extremely efficiently armed. However, no lion would risk encountering the massed horns of a whole buffalo herd.

Right: large breeding colonies of birds form extremely efficient defensive communities. A large crowd of birds is undoubtedly more of a deterrent to an egg robber than the self-protective efforts of a single breeding pair.

Below: one of the best known group defences is the circle of musk-oxen. Neither bears nor wolves have a chance of reaching the well-protected young in the centre.

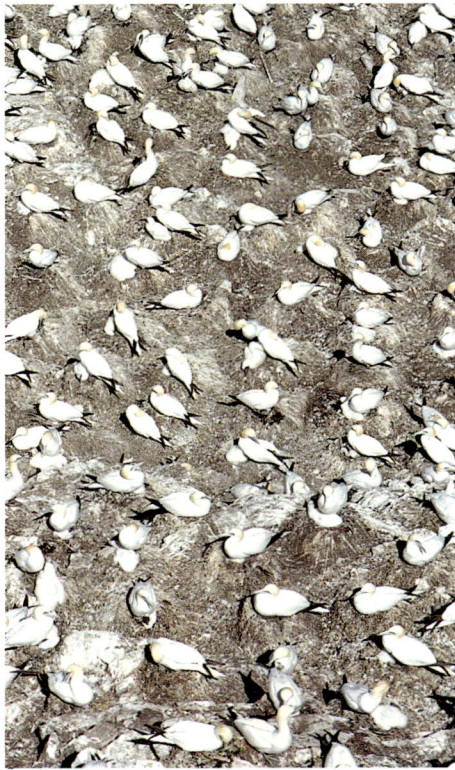

Many animals which might be overpowered by predators when they are on their own are transformed into formidable opponents if they are in a group. The best known examples among larger animals are perhaps the protective circles of musk-oxen whose rigid phalanx of horns will keep any wolf at bay. Herds of zebra also form a living barricade to ward off aggressive hyaenas. Small birds will often come together to mob a predator, chasing it away. Certain insects, too, provide a miniature example of this behaviour pattern. When threatened by ants, lacewing butterfly larvae position themselves so that they confront the enemy with a long line of sharp-pointed mandibles.

Another rather different but

Right: these tiny lacewing larvae, all hatched from a single clutch, form a hedgehog-like defence against an attacker.

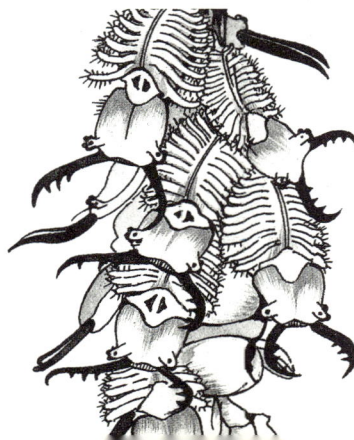

very efficient group defence mechanism is found in social insect colonies. Termites and certain ants have their own soldier castes, whose only task is to repel enemies. The mouthparts of many termite soldiers are adapted exclusively for

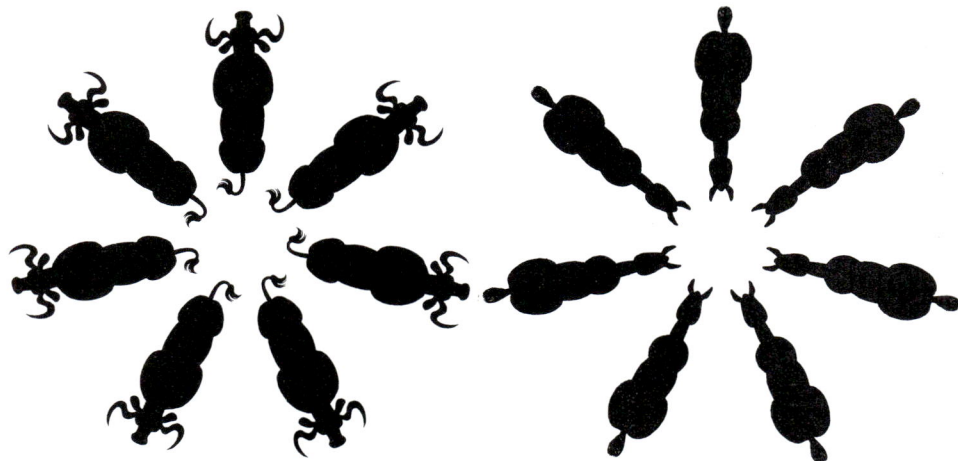

fighting and they have to be fed by their next companions. Like ants and bees, termites produce alarm substances to be used in self-defence. These are chemicals which are sprayed into the air by individuals under attack, so that in a short time the whole colony becomes a cohesive unit with thousands of heads and stings waiting to repel the intruder. Colonies of aggressive wild bees can inflict severe injury or even kill humans.

A comparison between the defensive formation of horned animals and horses. Both direct their weapons, horns or hooves, outwards against the enemy.

DISSIMILAR PARTNERS

Left: the small crustacean (Alpheus djiboutensis) digs the shared burrow while the goby (Cryptocentrus) watches the surroundings and warns its almost blind partner of danger.

Above: the thorny branches of an acacia grow bladder-like hollows which ants use as homes. In return they clean the host plant's surroundings of competing weeds and ward off plant-eating animals.

Below: chattering and calling, the honey-guide (Indicator) alerts the honey badger (Mellivora) to the presence of wild bee colonies. When the badger has opened and raided the bee nest, there is usually enough food left over for the bird.

The word symbiosis is derived from the Greek, and it means living together. By this biologists mean a living association of two organisms which reciprocally profit from each other. There are many examples in the plant kingdom. For a long time lichens were thought to be independent organisms, but actually they represent a close symbiosis between algae which can harness the sunlight photosynthetically and fungi which are capable of releasing mineral salts that are to be found in hard rocky substrates.

There are also other forms of close symbiosis and where both share food it is known as commensalism. But often one partner is involved in a protective role, the other looking after the diet. Hermit-crabs attach poisonous sea-anemones to the snail shell in which they live; the anemones feed on scraps from the crab's food. The position is different in the case of ants which live in the thorns and bladder-like enlargements of acacia twigs. Here too the plants provide protection, and also space in which the ants can live. This type of reciprocal service is extremely common and in some cases the benefit received from one partner is not always the same as that received by the other. There are numerous transitional stages between true symbiosis and parasitism.

Left: the calcareous skeleton of the small coral (Heteropsamnia cochlea) *grows round the worm* Aspidosiphon. *Its stinging arms protect the worm which drags the otherwise immobile coral around.*

The stinging arms of sea-anemones are feared and are normally lethal to small fishes. Only the little anemone-fish Amphiprion *can move about safely among them.*

SERVICES RENDERED

Below: Herodotus and Aristotle recorded that the Egyptian plover (Pluvianus aegyptius) flies into the jaws of a crocodile to remove leeches and food remnants. In more recent times, doubt has been cast on this old and much repeated story. So far no scientific investigation has conclusively verified this pattern of behaviour.

Some symbiotic relationships between animals are concerned with cleanliness and hygiene. This is the case with certain birds and large African animals such as buffaloes, rhinoceroses and giraffes. The oxpecker will hop about on the animal's back, flanks or belly with the agility of a woodpecker on a

African bird which does a similar service is the cattle egret or buff-backed heron. It too pecks parasites from the hide of ungulates and warns other herbivores of impending danger, rising into the air with large, flapping white wings.

There are similar associations in the warm, tropical seas. Cleaner

Right: a cleaner wrasse (Labroides) rids its client of parasites. In one experiment, removal of all cleaner fishes from a coral reef resulted in its rapid depopulation. After two weeks, the fishes which had not swum away showed skin ailments and abscesses due to the lack of cleaning.

tree trunk, looking for skin parasites, insects and other invertebrates. This is not always a delicate operation. When an oxpecker opens the hard lumps or warbles on a rhino's back in order to reach the parasitic warble fly larvae, it may draw blood and even drink it. Despite the pain this must cause, the bird is tolerated for in addition to removing the parasites, it acts as a sentinel. Another

wrasse take up fixed stations, and might be said to run regular cleaning businesses. In a single hour one of these small, longitudinally striped fishes may relieve up to fifty clients of their parasites. In doing so they even swim in under the gill-covers and into the larger fishes' jaws. In submarine caves their role is taken over by small red and white cleaner prawns. As for sharks, they permit an even more

Left: ants of the species **Lasius fuliginosus** *with the aphids with which they form a symbiotic partnership. The white areas on the backs of the aphids are the secretions of their wax glands. To a certain extent the wax protects the aphids from predators.*

permanent type of association for the pilotfishes that cleanse them are their constant companions.

The partnership between ants and aphids is likewise advantageous to both sides. The ants protect the aphids from enemies, such as ladybird larvae. In exchange, the aphids allow themselves to be 'milked'. The ants touch the aphids with their antennae and the aphids then lift up the rear end of the body and release a drop of sugary liquid. This procedure is similar to that of an ant begging for food from a nest companion. Some ant species keep aphids in their nest, 'employing' them to suck sap from underground tree roots. In others the relationship between ants and aphids is so close that the queen takes some aphids with her on her nuptial flight.

Above: along with the ox-pecker, the cattle egret performs the vital function of ridding large ungulates of parasites. The expansion of domestic cattle raising gave these birds a new lease of life, and they are nowadays one of the commonest African species. About forty years ago the cattle egrets also reached South America and Australia.

119

TERMITE COLONIES

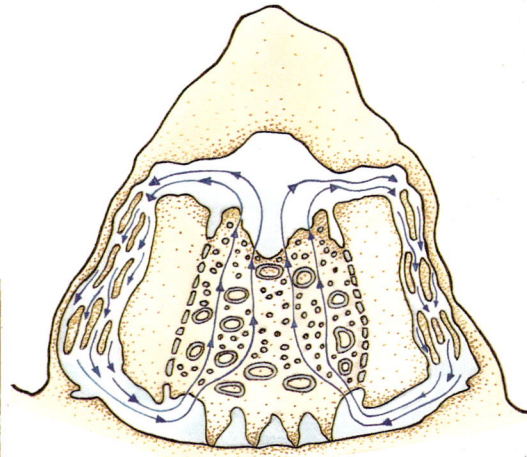

Termites probably evolved originally from social cockroaches. In common with these, termites have the ability to digest cellulose with the help of symbiotic micro-organisms. While some of the species live within tree trunks and feed on wood, most of the others, including the higher termites, live on dead and decaying vegetation, and some cultivate fungi in underground chambers within mounds of earth that are sometimes built to a height of five to six metres. They are, in fact, animal architects without parallel.

A colony of higher termites may contain several million individuals. Most termites are eyeless but have a highly specialized sense of smell and touch. If a termitarium is struck with a stick, the answer from inside is a strange rippling noise. This is produced by

Above: confusion gives place to discipline when a grass stem has to be removed. Termites do not understand one another by visual signals nor by drill commands. Most of the insects are blind and cannot perceive aerial sounds. Their language consists of scent signals secreted from glands and interpreted by means of the delicate scent organs on their antennae.

Right: a young termite king with attendants.

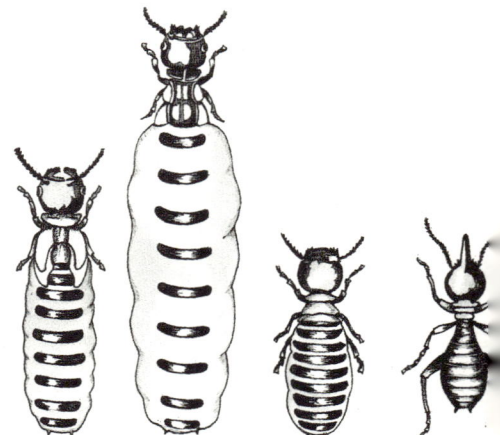

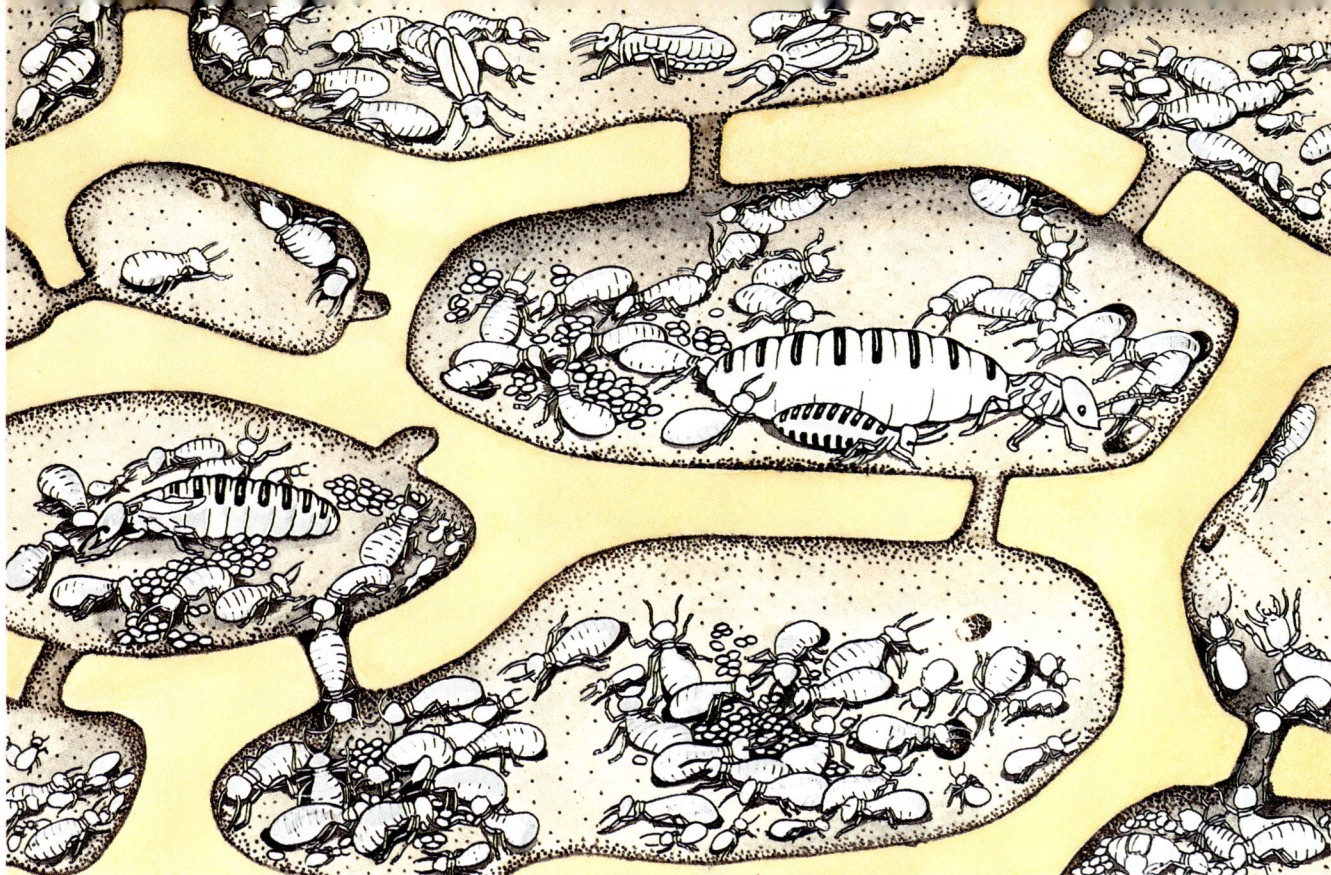

several thousand soldier termites beating their hard heads on the gallery walls, thus sounding the alarm. However, the main part of a termite colony consists not of soldiers but of workers. They have soft bodies and are wingless, responsible for building the nest, caring for eggs and larvae, and obtaining food for the soldiers and reproductive termites. The winged reproductives swarm out, shortly after being born, at the start of the rainy season. The majority of these are consumed by birds but a few manage to mate and establish a new colony, their wings dropping off after the nuptial flight. From then on the partners remain as king and queen of a growing community; from time to time the king mates again with the queen. In some species the queen increases her weight about 125 times. She becomes a monstrous egg-laying machine, immobile in her cell, continuously fed by the workers. For several years the queen lays up to thirty eggs per minute, or thirteen million per year.

Drawings above: termitaria are not simply mounds of earth, but highly organized structures. The arrangement of the air channels (left) guarantees that the used air, warmed by the metabolism of innumerable insects, rises, flows along the outer walls of the structure, cools down and sinks. In this way it gives off carbon dioxide and takes up fresh oxygen. Thus regenerated, it flows back downwards into the centre and up into the living quarters. The circulation in this wonderful "lung" is unique and driven only by the metabolic warmth of the inhabitants.

The caste system of termites is different from that of ants and bees. In many species the workers represent transitional stages of development, whereas the soldiers and sexual individuals are final products. The drawings at the bottom of this and the facing page show, from left to right: secondary queen, primary queen, tertiary queen, nasute soldier, jawed soldier, king and worker.

ANIMAL LANGUAGE

Facial expression in "huuh" call

Open full grin (when excited)

Playful face

Sulky face

Closed full grin (when puzzled)

The possibilities and limitations of animal language are best appreciated by comparing it with human speech. The latter has a vocabulary of many thousands of words which can be combined in an endless variety of different ways to form sentences. We are capable of giving a name to anything. We can expand our vocabulary to describe new inventions and express new concepts. The fact that we are not born with a built-in understanding of words is not necessarily a disadvantage, because it gives human speech flexibility to adapt to any new situation.

The abilities of animals to communicate with one another are, by comparison, more inflexible and considerably more restricted. But they are not limited to sounds alone. Indeed, they can call upon a wide range of other sensory means of communication, notably smell, taste and touch. However they send such signals, they either express mood or relate to defined situations that are often vital to their very survival. Such messages may be translated as a threat, a greeting, a desire for food, a danger warning, and so forth, and although we cannot interpret more than the most obvious, they are clearly adequate for the purposes of the animals concerned. Probably each individual species has fewer than fifty signals at its disposal. The higher the species,

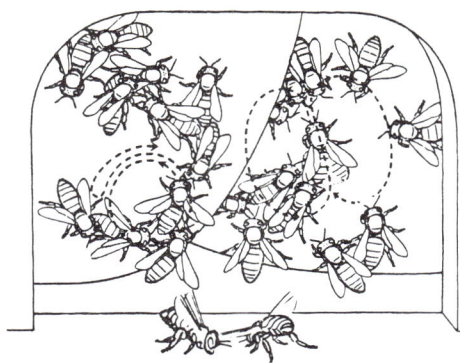

Below and right: the dance language of bees assumes two forms. The round dance (on left of picture) signifies a food source close to the hive. The tail-wagging dance (right) gives information on the direction and distance of a food source farther away.

Below: in the tail-wagging dance, the foraging bee who has discovered the food performs a figure of eight on the comb. The central connecting section, characterized by vigorous wagging of the abdomen, is of particular importance. The speed with

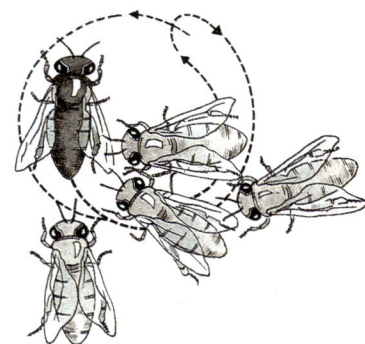

which the dancer performs the figure tells the others how far away is the food.

the more complex are its activities and the more elaborate the methods of communication and response. Wolves and anthropoid apes are a case in point.

Comprehension is particularly stereotyped in insects and other invertebrate animals. To a given signal there is usually only one possible answer from the partner. Nevertheless, insects have one of the most precise forms of communication in the whole animal kingdom. Homecoming bees, which have foraged successfully for food, are able, by dancing, to tell their hive mates the exact direction and distance of the food source. They perform a kind of figure of eight on the vertical surface of the comb,

eagerly watched by their companions. The most important element of the whole dance is the middle piece between the two loops: the tail-wagging section, so called because the bees actively wag their abdomens. The direction in which this portion of the dance is performed gives information on the locality of the food source. If the wagging is upwards it means that the food lies in the direction of the sun; wagging downwards means that with the hive as base, the food lies in exactly the opposite direction to the sun. Directional information is also contained in the dance. The faster the tail wagging is performed, the closer the food source is to the hive.

The direction in which the tail is wagged indicates the whereabouts of the food in relation to the position of the sun. The assembled bees which need to know this location follow the dance attentively, using their antennae to maintain contact with the dancer.

123

FISH SHOALS

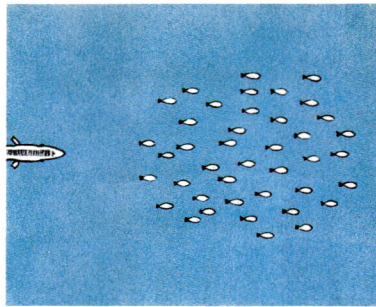

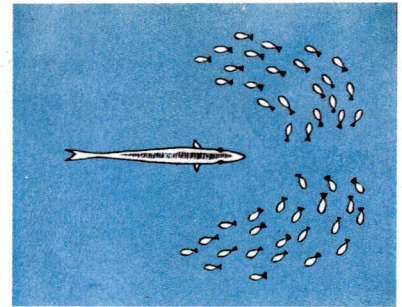

Many small fishes that live in shallow seas form shoals which often contain millions of individuals. Herrings and sardines are typical examples of such shoaling fishes. It is because of this tendency that modern fisheries have developed as massive industrial enterprises.

Fish shoals are not only aggregations but highly organized structures which in certain situations behave as a single organism. It is a remarkable fact that individuals in a typical shoal are roughly the same size and swim more or less parallel to one another. Sight and the lateral line organ indicate to each fish the exact position and movements of its neighbours in the shoal. The lateral line organ usually consists of a sunken groove of mucous pores extending along each side of the body from head to tail. It contains sensory cells which register every slight change in water pressure. This system of touch at a remove enables the entire shoal to execute rapid and complicated manoeuvres with amazing coordination – as if performed by a single fish.

There has been much discussion about the importance of fish shoals as protection against predators. However, two points appear to be clear. The first is that shoaling fishes tend to have fewer contacts with predators than would be the case were a similar number of individuals to be distributed separately throughout the surrounding water. In an encounter with a

Above: large number of rock crabs (Grapsus) *live in the intertidal zone of the Galapagos Islands. Each individual searches for its own food and scarcely notices its fellow crabs.*

Right: mixed colonies of birds are often formed, mainly due to lack of space. The birds cooperate against attackers. Pelicans and boobies are seen here on a guano island off the South American coast.

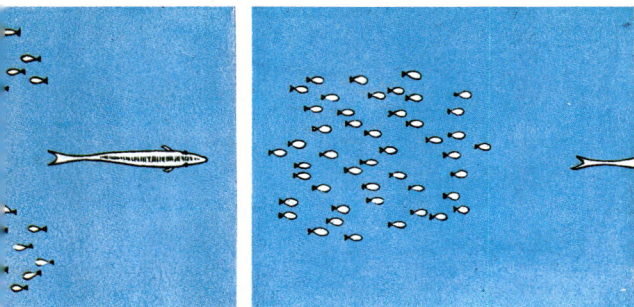

Left: fish shoals are able to perform astonishingly complex manoeuvres. Faced by an attacking predator, they can disperse in all directions, outflanking it and then reuniting as a shoal.

Below: fishes in a shoal swim parallel to one another and react by reflex to the movements of their neighbours, so that the shoal operates like a single organism.

predator a single prey fish would immediately be lost, whereas in a shoal, only a comparatively small number of fishes will be eaten. The second point is that by well coordinated actions a shoal can often confuse a predator. It can divide and disperse in front of an attacker and then reunite behind it. Nevertheless, such tactics are not always successful: in the course of lightning-swift attacks, tunny and barracudas can cut whole swathes in a shoal of small fishes. In general, however, shoaling behaviour greatly improves a fish's chance of survival.

Left: schools of dolphins cannot be compared with large anonymous fish shoals. Here the animals know one another personally, and there is a complex net of social relationships. The drawing shows a group of dolphins (Delphinus delphis) supporting a wounded colleague so that it can breathe air (after Pilleri).

125

SOCIAL MAMMALS

Right: the marching order of baboons. In the centre, and therefore best protected, are the females with their young. Close to them are the high-ranking males. Towards the outside are lower-ranking males and beyond them the juveniles, who are the first to fall victims to a predator.

Above: impala herds are harems, groups of 15–20 females held together by a buck. Impala bucks defend their harem, not a specific territory.

The colonies of social insects such as bees, ants and termites are highly developed but anonymous, in the sense that the constituent members lack individuality. The social communities of certain mammals are also well organized. Dolphin schools, elephant herds, baboon troops, wolf packs and lion prides have all attained a degree of mutual assistance and interdependence that is not found elsewhere in the animal kingdom. Naturally, our closest relations, the large anthropoid apes, belong to this category. Unlike the constituents of a fish shoal or an insect colony, the members of these mammal communities have distinct personalities and are fully aware of one another as individuals. Thanks to their relatively large brain, they acquire experience and understanding of the various laws and codes of behaviour of the group. Many communities are governed by hierarchies, and every detail relating to social rank is gradually stored within the memory of each

Right: mother baboons enjoy a high social status. However, the position of a female depends on whether she mates with a high-ranking or with a subordinate male. When two females form an alliance, the one who was inferior is raised in rank; such pairs, who perform varying services for each other, can climb quite high on the social ladder.

126

Left: dwarf mongooses (Helogale parvula) live in a matriarchy. The senior animal in the large family is the alpha female (1), who alone can breed with the alpha male (2), subordinate to her. The father joins the young (3) in fighting against other mongoose families, while subordinate adult males (5) are responsible for dealing with other hostile species. Adult females (6) are also subordinate. Young males (4) function as watchmen (after A. Resa).

group member. Any change of rank resulting from combat between rival males is automatically assimilated by all animals.

Among mammals, the basis of social behaviour is apparently the need to suckle the young. So the mother-child relationship provides the original stimulus for all such social groups. When the young of various litters remain together for several years, a kind of family community is established. With lions it is clearly the females who form the backbone of the community. Daughters remain with the mother, while young males move away and fight to join another pride. In some monkey communities the female sets the tone; in others there are separate hierarchies for males and females.

Female mammals tend to spend more time with their offspring than with the males – a tendency which helps to reduce the birthrate.

Above: elephant herds form a very close social community. The animals support one another in all situations and rear the calves communally.

Below: prairie dogs (Cynomys) live in immense colonies.

PARASITES AS PARTNERS

Above: the trematode Leucochloridium *uses a* Succinea *snail as an intermediate host. It penetrates the snail's tentacles. When birds consume what they take to be insect larvae, the parasite can complete its development inside them.*

Above, centre: the wasp Odynerus *is closing its nest entrance. It has scarcely flown away to fetch more mud, when the parasitic hoverfly appears (above, right). The latter lays its egg in the nest while still airborne. Its larva will destroy the wasp larva.*

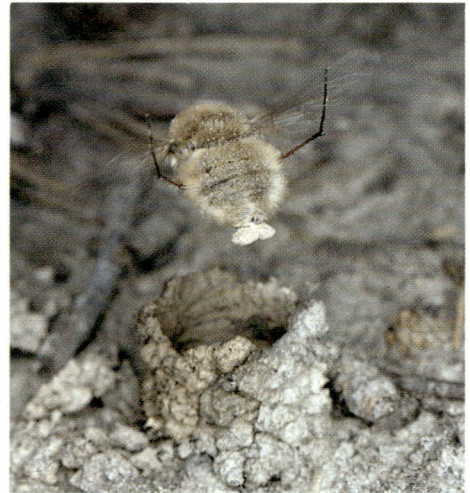

There is only a thin line between predation and parasitism. In general, a predator kills its prey, whereas a parasite exploits its host without killing it. However, there are numerous transitional forms. The larvae of certain carnivorous flies, for example, assault their victims by initially eating their way into the fatty body parts of moths, thus doing comparatively little damage, and later penetrating the vital organs, in this way killing the host. Such mixed forms,

which start by behaving simply as parasites and then, just before they are transformed into nymphs, become mortal aggressors, are known as parasitoids. A true parasite, however, does everything to keep its host alive because if the latter dies, it will no longer be able to work on the parasite's behalf. Naturally, the body of the injured party puts up a resistance, either through its immune system against internal parasites or by scrupulous body cleansing against external parasites. Sometimes parasites are encapsulated and isolated in a restricted area of the host's body; yet astonishingly, the parasite and its victim often find it possible to tolerate each other. The host may, in time, profit from its parasites, establishing an equilibrium which benefits both sides, more akin to symbiosis. Probably, all cases of symbiosis originally evolved from parasitism. In any event, there are numerous transitions between these two forms of co-existence.

Right: having devoured the swallowtail pupa, the larva of the ichneumon fly itself pupates in the empty pupal case and emerges as an imago.

HOUSING ESTATES

Facing page: this woodpecker family has settled in the middle of the paper nest of a Crematogaster *ant. Although the guests may consume some of the ants, they are generally tolerated.*

The large communal nests of the sociable weaver last for several bird generations and are continually enlarged until one day the supporting branch breaks under the weight. With its dozens of entrances, this structure also offers protection to many other bird species.

Monitors often live in occupied termite mounds, which they prise open with their hard claws. It is not clear whether the termites derive any advantage from their subtenant.

Breeding colonies of the sand martin are located on steep cliffs. The birds live solitarily and each pair rears its own young. Now and again a few males may try to intrude but the resident male guards against this.

The nest colonies of the oryx weaver look something like the initial stage of the communal nests of the sociable weaver. As protection against egg robbers, the nests are built as far out on the branches as possible. However, this does not always deter snakes.

Blocks of flats and large housing estates are common features of urban life. In the animal kingdom too there are interesting examples of species, particularly birds, which live together. Thus, in Africa, the enormous communal nests of the sociable weaver (*Philetarius socius*) may measure up to five metres across. They actually consist of individual nests under a single grass roof. As a rule, twenty to thirty pairs of weavers live together, although constructions with 125 entrances have been found. Such buildings may provide generations of birds with protection and shelter. They may be further enlarged, until one day the supporting branch breaks, and the whole elaborate structure tumbles to the ground.

Not surprisingly, abandoned nest complexes often harbour squatters. In addition to an army of parasites, other birds frequently move in: small parrots and pygmy falcons often take advantage of empty nest chambers. Sparrows and starlings, too, will build their nests among the branches of the much larger nests of eagles and storks, thus forming communities. In this way the small birds benefit additionally from the protection provided for them by their larger neighbours.

The accommodation of small animals is likewise often shared or borrowed by larger subtenants. In Africa, certain birds that nest in cavities take up residence inside termite colonies, so enjoying the protection of the insects with their self-defensive instincts. But those woodpeckers which invade the paper nests of *Crematogaster* ants show less gratitude to their fellow tenants by eating some of the insects every day.

MIGRATION

Migration is one of the most basic and astonishing phenomena of animal existence. Many birds breed in temperate or cold latitudes because in summer the conditions there are very favourable for rearing the young, with plenty of food available. If they remained there throughout the winter, however, they would starve. So they move southwards, perhaps for create a white wilderness and when drought parches the waterholes. Fish follow sea currents to find plankton and herbivorous mammals relinquish overgrazed steppes in search of new pastures. They travel together and on arrival at their destination disperse over wide areas so that each individual can find sufficient food. For some species of birds, however, this

Facing page: in autumn the North European brambling flies to central Europe, sometimes forming incredibly large flocks. For instance, in the winter of 1977–78 a gigantic colony settled in a valley near Basel, Switzerland, darkening the sky every evening. Their numbers were estimated at 30 million.

Right: various grazing animals undertake extensive migrations. These include the wildebeeste of the Serengeti Plain in Tanzania. Surprisingly, however, the wildebeeste that live in the Ngorongoro crater, only a few kilometres from the Serengeti, do not undertake seasonal migrations. They usually move about within the crater, but with no definite rhythm.

thousands of kilometres, in order to survive the cold. Observations and ringing techniques show that some species spend half the year travelling to and fro.

The principal reason why animals migrate, however, is to find enough food to survive during the season when their usual habitats become climatically unendurable, when ice and snow strategy, originally a recipe for survival, has become increasingly dangerous. The flight paths of certain European migratory birds traverse Italy, where thousands of spare-time hunters, predicting their direction and arrival time, decimate the flocks for so-called sporting purposes. In Asia some rare species of crane are similarly threatened when on migration.

WANDERING FLOCKS AND HERDS

In spring, when the night temperature reaches about 5°C, toads wake up and often wander quite a distance from their winter quarters to the waters where they were hatched, in order to spawn there. It is still not clear how they find their way.

American pelicans do not migrate but sometimes travel great distances in search of good fishing grounds. They adopt a wedge formation in flight.

There are some 200,000 zebras living on the Serengeti Plain; like the wildebeeste, they too wander in search of food.

Most bats of the temperate zone hibernate. To reach their winter quarters, where they sometimes congregate in vast, densely packed masses, they often fly many hundreds of kilometres.

The Greek naturalist Aristotle was the first to consider animal migration on a scientific basis. Although he was incorrect in assuming that pigeons, starlings, swallows and some other birds hibernated in winter, he was right in stating that cranes came to the Nile to spend the winter, and he described the journeyings of pelicans. Two thousand years later there are still many unsolved problems in the field of migratory behaviour, but it has been verified that most animal species move or wander at certain times of year. Nowadays true migrations are taken to mean those animal travels from one well-defined region to another, with a specific purpose in view, notably for overwintering, breeding and seeking a richer food supply. Apart from these, there are those apparently aimless movements, which seem to be initiated by the simple urge to explore. The result of such population shifts is to increase the distribution range of the animal species concerned. More or less all classes of animal contain forms which wander as

true migrants. Innumerable tiny organisms, known as zooplankton, travel almost passively with the sea currents, on fixed routes over thousands of kilometres. A very valuable source of food, they influence the migrations of fishes, sea-birds, marine animals and marine turtles.

Among the insects there are

mass migrations of temperate zone butterflies and the irruptions of migratory locusts.

Numerous amphibians distributed over wide areas live as solitary individuals but at breeding time return to the waters in which they were hatched or born.

Many hoofed animals under-

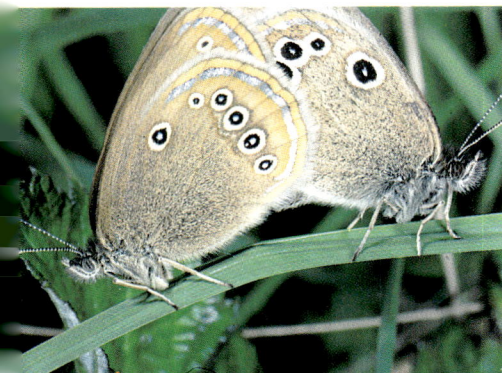

take seasonal migrations. In autumn, mountain animals move into the valleys, northern herds migrate south in search of more favourable conditions, while grazing animals of the tropics endeavour to avoid periods of drought. Every time they set off on their long excursions, they attract numerous predators in their wake.

THE BIBLICAL SWARMS

Facing page: a swarm of migratory locusts darkens the sky in western India. With the help of the wind, such swarms of millions may migrate for hundreds of kilometres, causing extensive damage to the vegetation, sometimes leading to famine.

The five species of lemming live in the Far North. Under favourable climatic conditions, some of them reproduce explosively and move off on mass migrations.

The various species of lemming are the most numerous vertebrate animals in the subarctic. Best known, because of its mass migrations, is the Norway lemming.

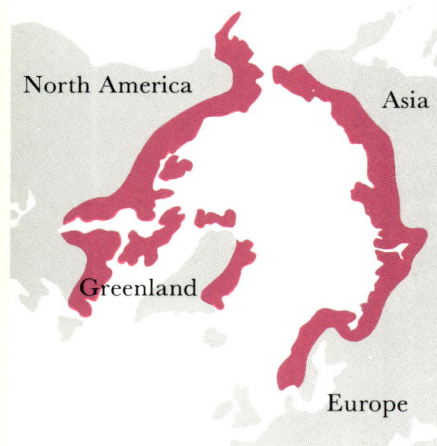

There are seven species of locust – grasshoppers which occur in open country, in tropical and subtropical areas. In Africa, where the populations are largest, the desert locust is particularly feared; this is the species which forms the enormous swarms mentioned in the Bible.

These insects do not always develop into migrators or wanderers. If they are hatched in areas with rich vegetation, they assume striking green coloration in the flightless juvenile or hopper stage and also as adults. These are the so-called solitary locusts, which behave much like other grasshoppers. They spend the day hidden and move out at night individually in search of food. If, however, they hatch in desert areas or in places characterized by long periods of drought, both hoppers and adults take on a pattern of bright yellow and black, and their behaviour is completely different. For a long time the two phases of this locust were regarded as different species. Already in the hopper stage these 'wanderers' gather together in hordes. After a few weeks, when they are able to fly, they move away in search of wet areas because only there can they successfully lay their eggs. These swarms can be so enormous that they darken the sky like an eclipse of the sun, and the noise of their beating wings is almost unbearable. In zones where they land they devour all the vegetation. In a few minutes they can completely destroy the harvests. Migrating swarms cover a vast area, soar to great heights and travel thousands of miles. Irruptions of desert locusts are difficult to control, even by spraying.

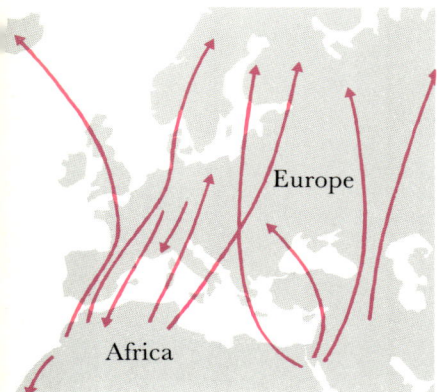

Right: in spring and summer the lesser tortoiseshell migrates north, sometimes in enormous numbers, returning south in autumn. It may travel more than 2000km and fly over high mountains. The map above shows the migrations of the European form.

THE PIGEON PUZZLE

Homing pigeons are derived from rock doves which nowadays still live wild in Scotland and the Mediterranean area. Rock doves nest in cliff cavities and fly several kilometres every day to seek food in the surrounding fields. Because they need to find their way home, they have a good ability for orientation. Yet the homing ability of wild rock doves is as nothing compared with that of trained pigeons. The maximum distance they can find their way home is about fifteen kilometres. If they are taken more than eighty kilometres away they are lost for they are not migratory birds. The homing pigeon has been selectively bred for that very purpose. Amazingly, a trained pigeon is capable of returning to its loft from a vast distance, flying some one thousand kilometres in a straight line in a single day.

Since olden times homing pigeons have been used to carry messages, and as the centuries have passed, so their homing ability has been continuously improved. Modern homing pigeons are larger and their aerodynamic build is more efficient than that of their ancestors. Their brain is bigger and their sense organs more acute. Pigeon breeders race their birds regularly and continually increase the distances covered during the course of a season. Only those birds which perform best in these races will have a chance to be used for further breeding, to produce pigeons that are more and more efficient at high speeds and

over long distances. Nowadays, there are specialized breeding stocks for long-distance flights and these birds are able to find their way home from distances of up to about 1600 kilometres.

Yet although breeders have raised these insignificant-looking birds to perform such astonishing feats, the secret of their success has not been revealed. We still do not know exactly how a homing pigeon brought to an unknown point of release can find its way home over hundreds of kilometres of strange terrain. The investigations of Kramer and Schmidt-Koenig in the 1950s showed that pigeons are able to utilize the sun as a compass: they know the path which the sun travels each day and from this they can calculate, by means of their internal clock, the direction in which to fly. However, this is not a sufficient explanation for their performance, because a bird travelling over unknown country needs not only a compass but also a map. So before the pigeon can fly home with the aid of its direction finder, it must first establish its exact position.

The way in which the pigeon constructs its map is another mystery. There are plenty of theories, including the possibility of infrasounds reflected from ocean breakers on faraway coasts, scents wafted through the air over vast distances and so forth. None of these hypotheses has thus far led to a satisfactory explanation. Many investigators believe that during their journey from the pigeon loft

Homing pigeons maintain flight direction by navigating with the help of the sun. In the absence of the sun, the pigeons use the earth's magnetic field. Professor C. Walcott mounted small magnetic coils on the heads of his pigeons, thus disturbing the natural magnetic field. He then let them fly once in sunshine and once in bad weather. In sunshine their ability to reach home was unimpaired, but when the sky was overcast, they were totally disoriented. After the magnetic coils had been removed, the same pigeons found their way home, even in cloudy weather.

to the point of release, the birds somehow manage to discover where they are. The observations of the late William Keeton would appear to call into question the solar navigation theory. Keeton established that pigeons can find their way home even when the sky is overcast. Detailed experiments have shown that when the sun is completely obscured, the earth's magnetic field is used as a compass. Furthermore, Professor Walcott and his colleagues have discovered an organ which may be involved in magnetic orientation: between the roof of the skull and the brain there is a deposit of magnetite which has the tendency, very like a compass needle, to deflect the bird towards the north.

Below: it is still not clear how homing pigeons after release find their way back from a strange place. Professor F. Papi of Pisa University thinks that the sense of smell is involved, but American researchers have been unable to confirm this result. The picture shows a laboratory experiment on the perception of odours by pigeons.

Above: tests have been made with head cameras such as this to determine which external factors help pigeons to plan their flight routes. The developed pictures show that the sun plays an important role.

Birds display astonishing powers of migration. From their breeding grounds to their winter quarters they cover thousands of kilometres. Small birds usually fly at an altitude of 1000 to 1500 metres, but sometimes at above 4000 metres. Cranes have been sighted at 6000 metres, and bar-headed geese even at about 10,000 metres. Equally remarkable is the way in which, under hormonal influence, they control the timing of their migrations; so too is how they find the way. In 1890 the Danish ornithologist Mortensen fixed zinc rings to the legs of starlings, with a view to discovering where they had flown, but the paint he used

Bar-tailed godwit
(Limosa lapponica)

Golden plover
(Pluvialis apricaria)

Arctic tern
(Sterna paradisea)

Greenland

Icel.

North America

North Atlantic

South America

South Atla.

Ruddy shelduck
(Tadorna ferruginea)

Asia

North America

Pacific Ocean

Australia

for marking the wings wore off. Eight years later he used aluminium, scratching on a number and his address, and by 1899 he had ringed 164 starlings. Over 30 million birds have been ringed, and more than one million of these have been retrieved. Almost everything now known about the migration routes, destinations and travelling speeds of starlings stems from the pioneer work of Mortensen.

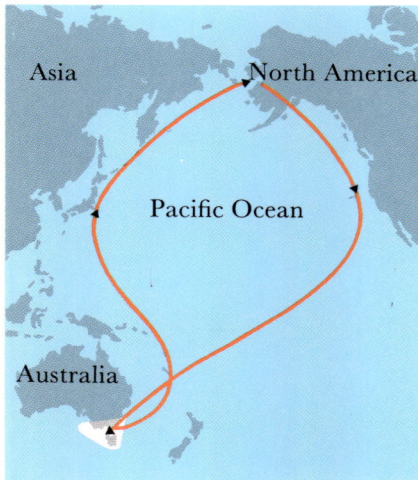

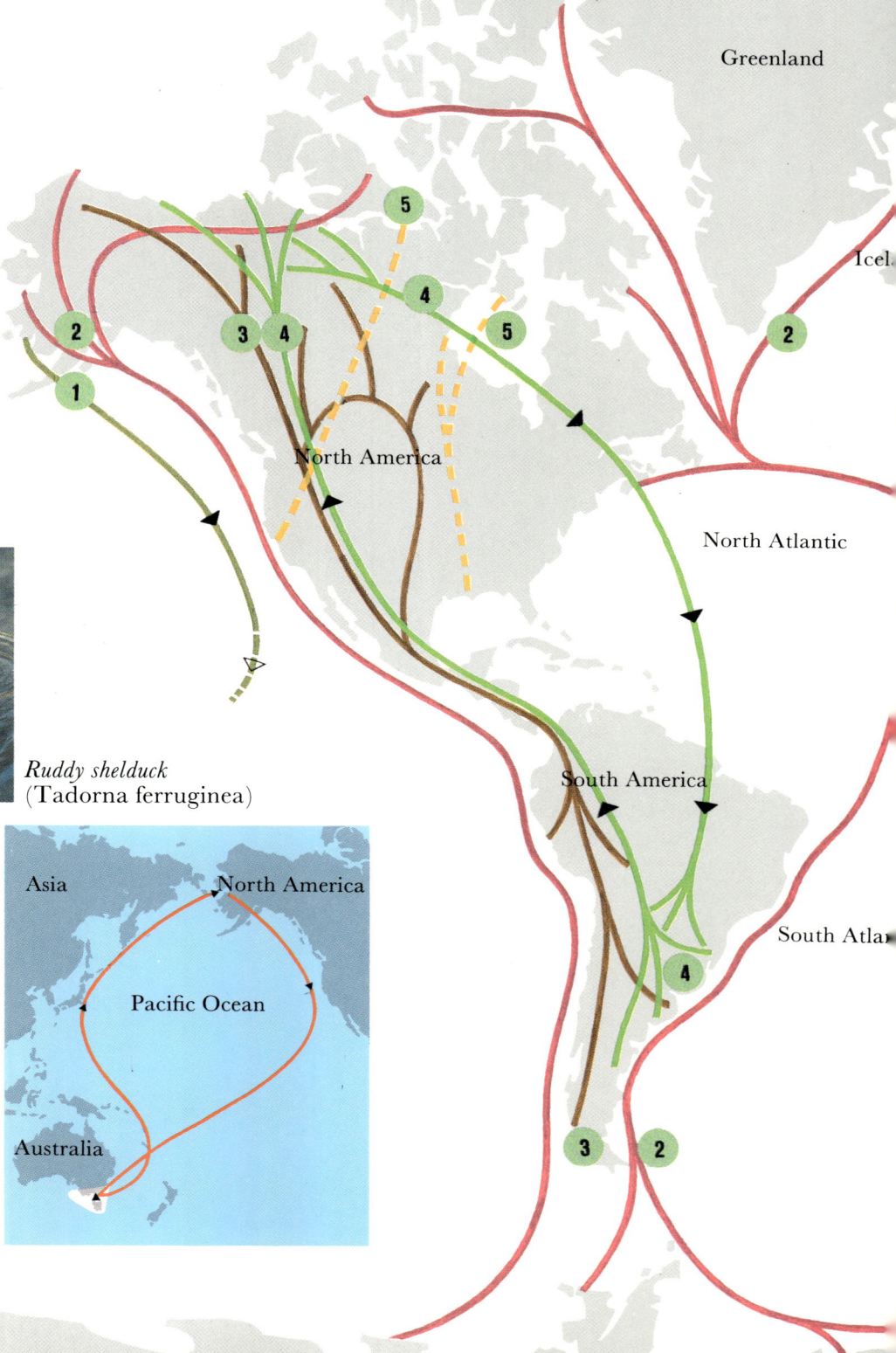

White stork
(Ciconia ciconia)

Swallow
(Hirundo rustica)

Pectoral sandpiper
(Calidris melanotos)

Migration routes:
1. (olive) Short-tailed shearwater
   (Puffinus tenuirostris)
2. (red) Arctic tern
   (Sterna paradisea)
3. (brown) Pectoral sandpiper
   (Calidris melanotos)
4. (green) American golden plover
   (Pluvialis dominica)
5. (orange, broken) Ross's goose
   (Anser rossi)
6. (pale blue) Swallow
   (Hirundo rustica)
7. (orange, unbroken) White stork
   (Ciconia ciconia)
8. (dark blue) Red-backed shrike
   (Lanius collurio)
9. (yellow) Bar-tailed godwit
   (Limosa lapponica)

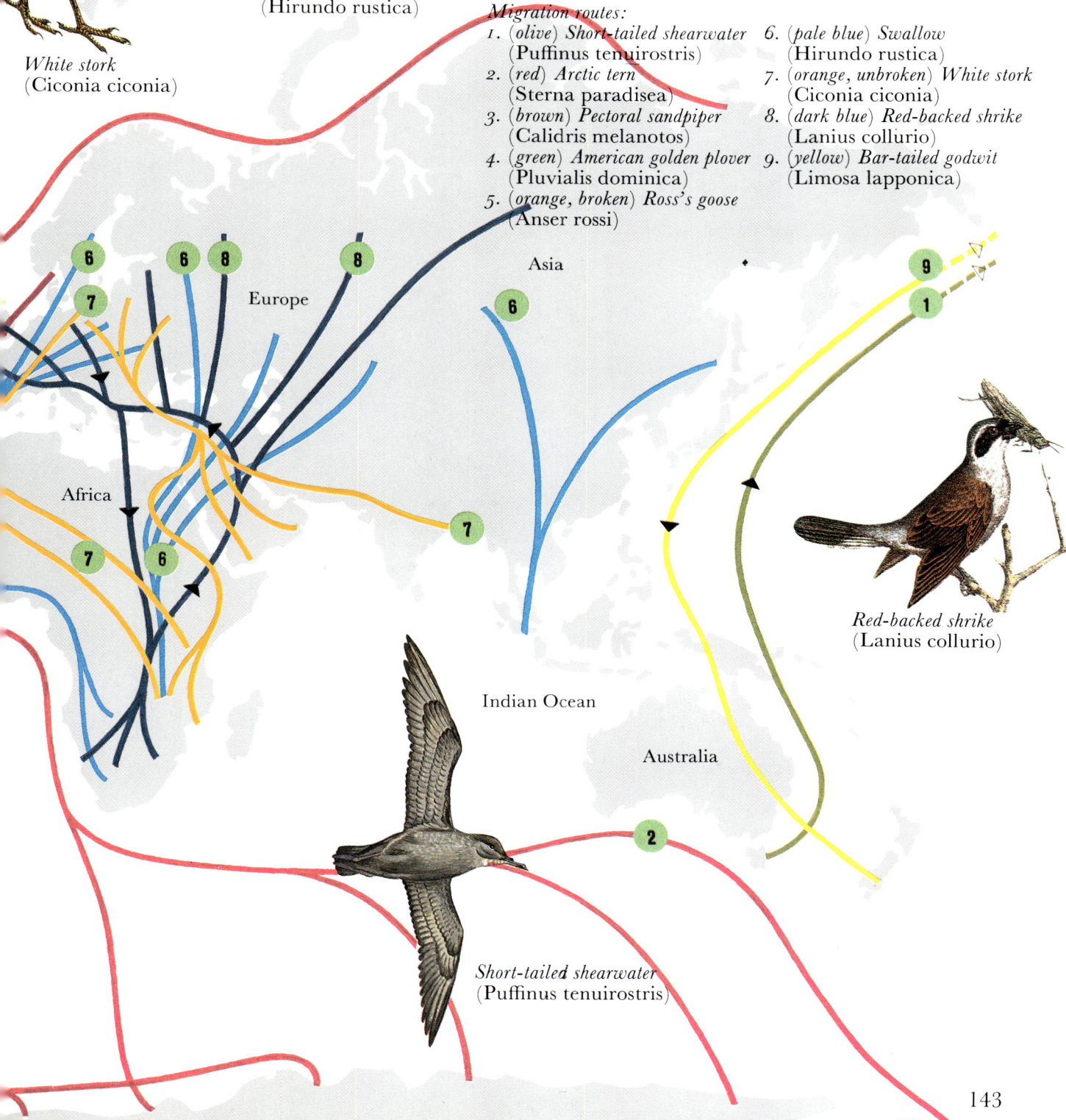

Europe

Asia

Africa

Indian Ocean

Australia

Red-backed shrike
(Lanius collurio)

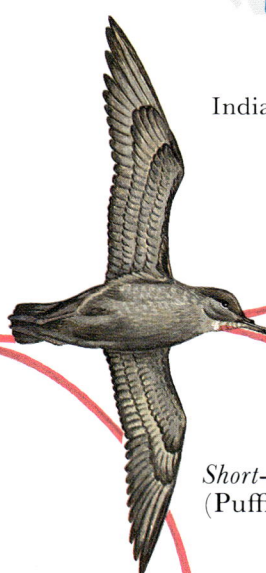

Short-tailed shearwater
(Puffinus tenuirostris)

143

There are various migrators among the marine mammals and reptiles. The migration routes are difficult to investigate but it is known that many whales, seals and marine turtles cover tens of thousands of kilometres annually. Sea otters regularly travel up to about a hundred kilometres, and even sea snakes swim long distances.

It is likely that all marine turtles migrate over long distances. Hawksbill turtles, for example, are known to spend part of the year off the Brazilian coast and travel more than 200 kilometres to lay eggs on Ascension Island.

Migration routes:
1. (*pale red*) *Hawksbill turtle* (Chelonia mydas), *which wanders between Ascension and Brazil*
2. (*blue*) *Fin whale* (Balaenoptera physalus)
3. (*dark red*) *Blue whale* (Balaenoptera musculus)
4. (*green*) *Walrus* (Odobenus rosmarus)
5. (*yellow*) *Fur seal female* (Callorhinus ursinus)
6. (*black*) *Fur seal male* (Callorhinus ursinus)
7. (*lilac*) *Grey whale* (Esrichtius gibbosus)

Hawksbill turtle (Chelonia mydas)

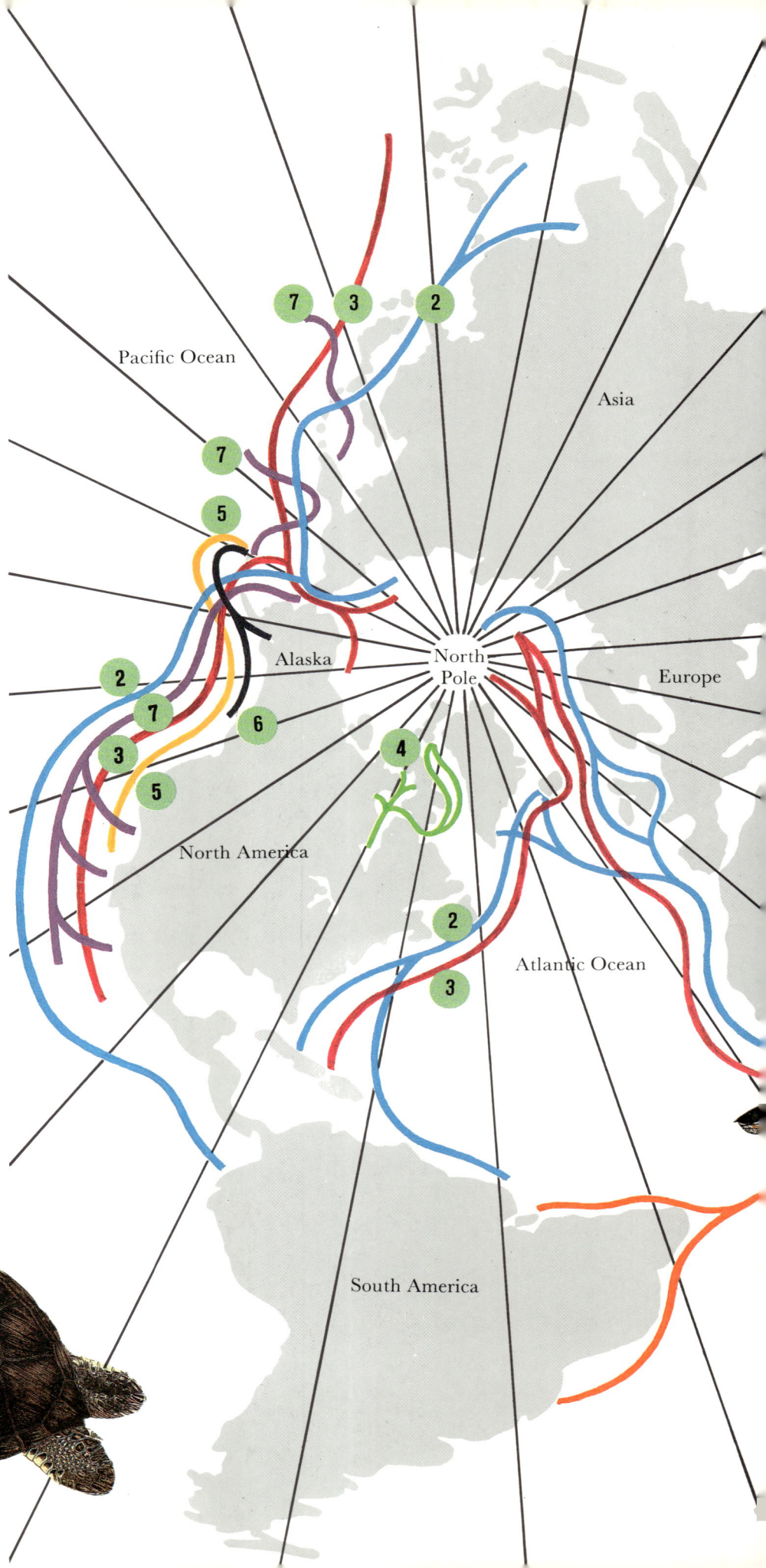

Pacific Ocean

Asia

Alaska

North Pole

Europe

North America

Atlantic Ocean

South America

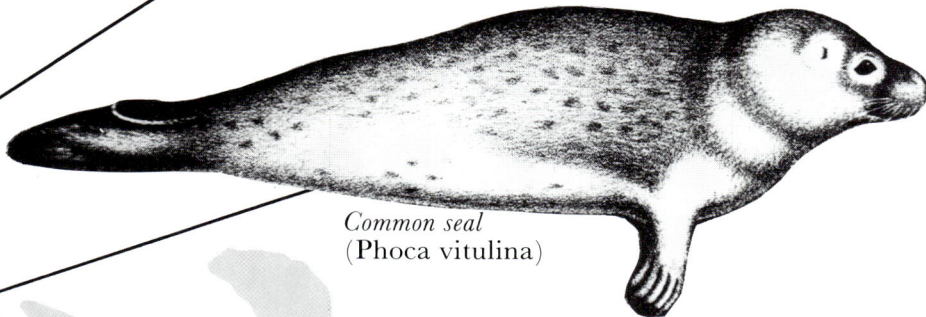

Common seal
(Phoca vitulina)

Walrus
(Odobenus rosmarus)

Africa

Greenland right whale
(Balaena mysticetus)

Greenland right whales, fin
whales and blue whales are
baleen whales which use the
sieving apparatus in the
mouth to filter mainly small
crustaceans (krill) from the
water. In autumn these

whales move to warmer
waters where the young are
born. They have to live on
their reserves because here the
krill population is only about
one-tenth as dense as in polar
seas.

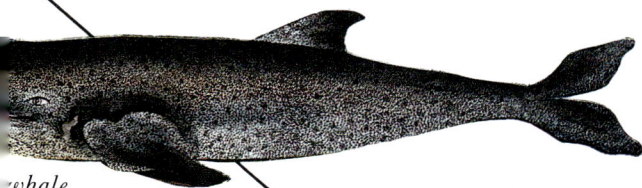

whale
aenoptera physalus)

Blue whale
(Balaenoptera musculus)

EELS AND SALMON

Migrating salmon encounter various hazards, such as brown bears, which show astonishing skill in catching these fast-moving fish.

Red salmon in their breeding area, in the shallow upper reaches of an Alaskan river. The males are usually silvery-grey, but at breeding time, which happens once in their life, the body is fire-red and the head green. At this time, for reasons unknown, the lower jaw develops an upward-directed hook.

Freshwater eels and salmon are among the most remarkable wanderers in the fish kingdom. Both spend a part of their existence in the sea and a part in rivers and streams, but their life-cycles are completely different.

American and European freshwater eels spawn in the Sargasso Sea, in the Caribbean area, at depths of 400 to 750 metres. After the larval stage the young eels of the American species swim to the east coast of America, while the European juveniles cross the Atlantic; by the third or fourth years of life they have reached the west coast of Europe, the Mediterranean and the Black Sea. Both species then swim up rivers where they spend several years before travelling back to the Sargasso Sea, where they spawn and die.

Eels are known as catadromous fishes, spawning in the sea and maturing in rivers. Salmon, which do precisely the opposite, are called anadromous fishes. Atlantic and Pacific salmon are hatched in the rivers of northern Europe and North America. After four weeks, when they have used up the contents of the yolk sac, the young salmon or alevins leave their hiding-places in the bottom gravel. When about ten centimetres long they start to swim downstream and into the sea. In the warmer, more southerly rivers this takes about a year, but in Alaska and northern Scandinavia about seven years. After approximately one to four years in the sea, the salmon move up the rivers again for breeding; at this time they do not feed, but live on their fat. After spawning, about 75 per cent of the salmon are so exhausted that they die, while the remainder migrate back into the sea.

Greenland

th America

Sargasso Sea

Atlantic Ocean

Europe

Africa

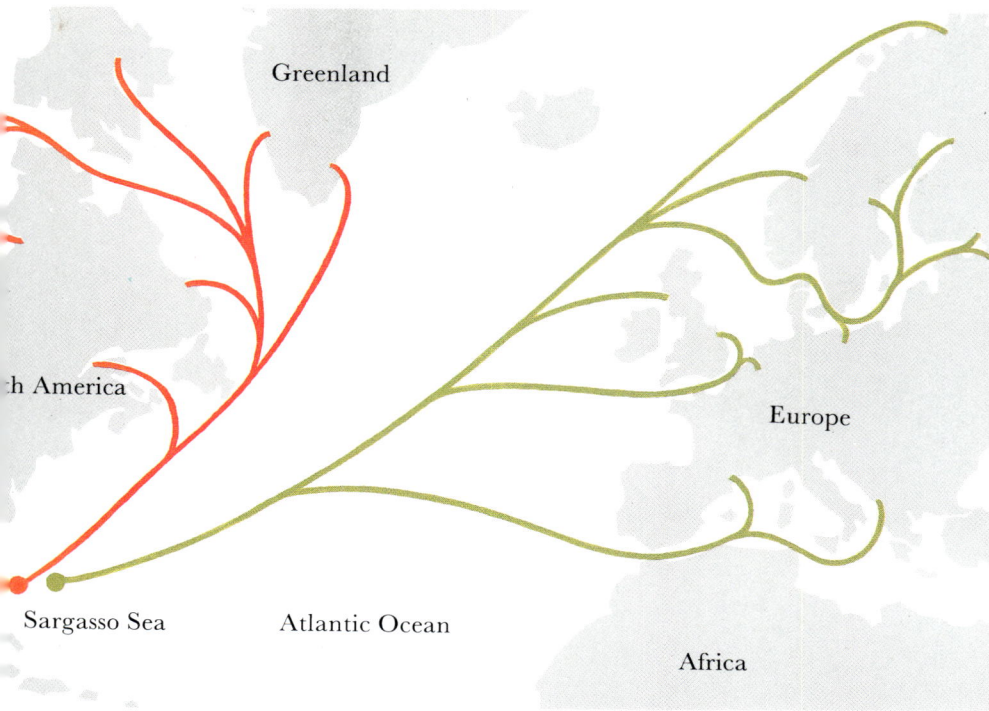

American and European freshwater eels hatch in the same, relatively small area of the Sargasso Sea. Here begins the long journey of the larvae, at first only a few millimetres long, to the river estuaries. The European eel covers more than 7000 kilometres. After about 18 months it is east of the Azores, about 7cm long, and has taken on a leaf-like shape. Metamorphosis into the slender glass-eel, about 15cm long, takes place in European estuaries. At this point the fishes, now three years old, start to swim up the rivers. Years later, as silver eels measuring 1m long, they undertake the return journey to spawn in the Sargasso Sea and die.

The larvae of the freshwater eel, shaped like a willow leaf, live in the sea. Before swimming up the rivers they become typically eel-shaped but are still transparent glass-eels or elvers. At the end of the first summer spent in the rivers, they develop into yellow and,

finally, silver eels. The diagram (below) shows the eel's life cycle from spawning through the various larval and juvenile stages to the sexually mature fishes which return to the Sargasso Sea to spawn.

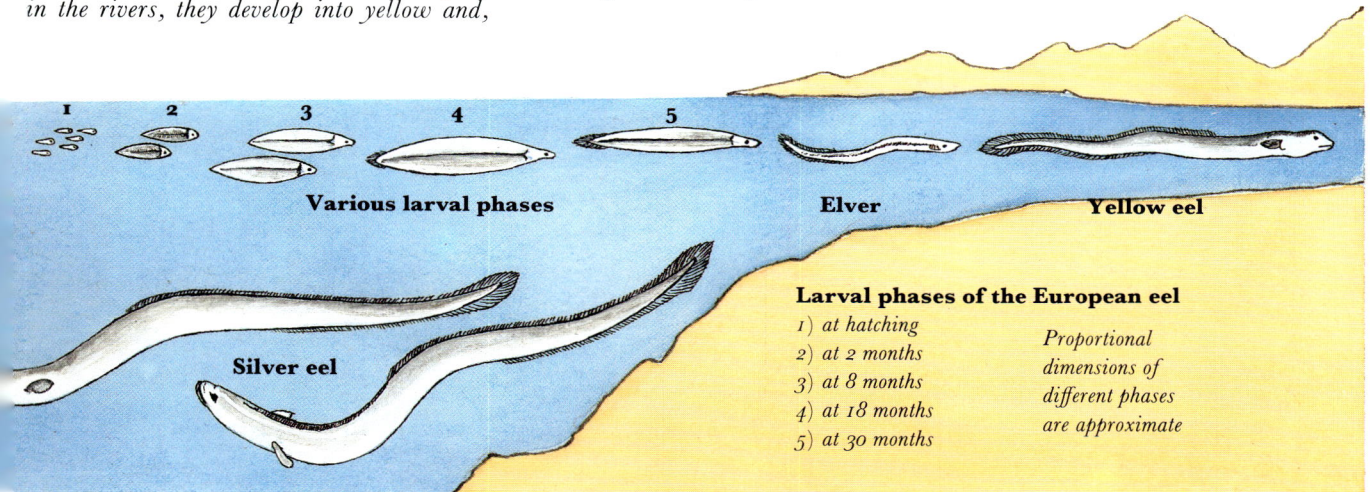

1   2   3   4   5

Various larval phases

Elver

Yellow eel

Silver eel

Larval phases of the European eel

1) at hatching
2) at 2 months
3) at 8 months
4) at 18 months
5) at 30 months

Proportional dimensions of different phases are approximate

Even waterfalls are no obstacle to eels. They bypass them by moving, snake-fashion, over land.

The eels remain for several years in peaceful inland waters, sometimes in stagnant ponds, until they are sexually mature.

Following pages: it is known that dolphins and other toothed whales migrate and that this depends, not on the temperature, but primarily on the density of fish shoals. However, only a few migration routes are yet known, and even those only partially.

147

THE ROAMING UNGULATES

Both the northern tundra and the grasslands of lower latitudes (prairie, steppes and savanna) support gigantic herds of hoofed animals or ungulates. However, these zones are subject to considerable climatic fluctuations: long periods of cold or drought alternate with fruitful summers or productive rainy seasons. Both these extreme situations make it necessary for the animals to migrate.

In prehistoric times saiga antelopes were distributed from the Baikal Sea to Britain. After they had been almost exterminated at the beginning of the 20th century, they have now, under strict protection, increased to about two million head. They spend the winter in the climatically favourable Caspian Sea region. In spring they migrate north for two weeks, covering some 350km, and give birth to their young. After that they move about 250km south-westwards to their summer grazing area and finally back to the Caspian Sea.

In the early 19th century one of the most spectacular sights of the North American prairie was still the mass migration of the herds of bison, commonly known as buffalo. These huge animals (the bulls are the largest of all American animals) were distributed from southern Canada to Mexico. The life style of the prairie Indians was based entirely on the bison: they followed the herds throughout the year, hunting them with bow and arrow, using their flesh for meat and their hide for clothing, but not decimating them. Around 1850 the bison population was estim-

ated at sixty million. In the following forty years the white settlers, as they built their railway from east to west, carried out a senseless mass slaughter of these animals, gunning them down for sport. Only about a thousand bison survived. At the last moment the species was saved from extinction by strict protective measures. Today there are about 30,000 head.

The springbok, an antelope which once roamed in gigantic herds throughout southern Africa, was another species that only just escaped extinction. In 1896 one observer saw a herd which was 20 kilometres in width and 200 kilometres in length; he estimated that there were over a million animals. A few decades later there were only a few surviving populations. With support from the government, the farmers had shot everything that was not a domesticated animal. In this way the bontebok, a large antelope, and the quagga, a subspecies of zebra, were finally exterminated. It is only in East

Africa that there are still large populations of herd animals. Here, enormous mass migrations can be observed in the Serengeti Plain. Each spring the herds move into the northern part and each autumn they return south. The main mass consists of about 350,000 wildebeeste (gnu), about 180,000 zebras and a few hundred thousand gazelles.

The wild horse, of which five subspecies once wandered through the steppes, tundra and forests of Eurasia, has disappeared; these were the ancestors of the domestic horse. Only a small population of Przewalski's horse survives. The European bison or wisent was almost exterminated.

It is only on the steppes of Asia that large migratory herds of hoofed animals still flourish. These are the saiga antelopes, vaguely resembling sheep, but smooth-skinned. This species was almost extinct, but thanks to official protection they now number almost two million. Saigas migrate twice yearly in herds of up to around 100,000, travelling about 350 kilometres in search of food.

In the Serengeti, wildebeeste, zebras and antelopes, comprising over a million animals, leave the open plains in the southern part of the National Park when these are affected by drought and no longer produce enough food. Up to the start of the rainy season, the animals live in the moister, bushy northern Serengeti.

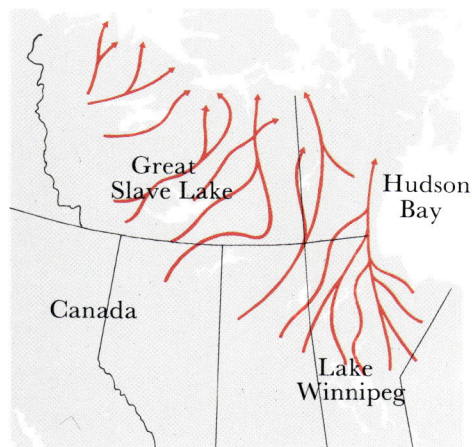

Caribou, Canadian reindeer, spend the summer in the open, arctic tundra. In autumn they travel, often several hundred kilometres, to the protection of the coniferous forests, sometimes forming herds of up to 100,000 animals. The map (left) shows their migrations in northern Canada.

151

PREHISTORIC MIGRATIONS

The various horse species evolved in North America during the Tertiary. Early on, Hyracotherium (Eohippus) reached Europe, either over the Bering land bridge or over Greenland. In the last few millions of years the modern horse species spread across the Old World (map, below). Nowadays the following, largely unchanged forms of wild horse still exist: the zebras in Africa, the wild asses (Asinus) in northern and north-eastern Africa, the half-asses (Hemionus) and Przewalski's horse in Asia, and the Exmoor pony in Europe. In North America, their original range, horse species became extinct, and were only reintroduced by the Spaniards.

Far right: the proboscidians (the drawing shows a mammoth) evolved in North Africa and later reached all continents except Australia and Antarctica (map, below).

Exmoor pony

Przewalski's horse

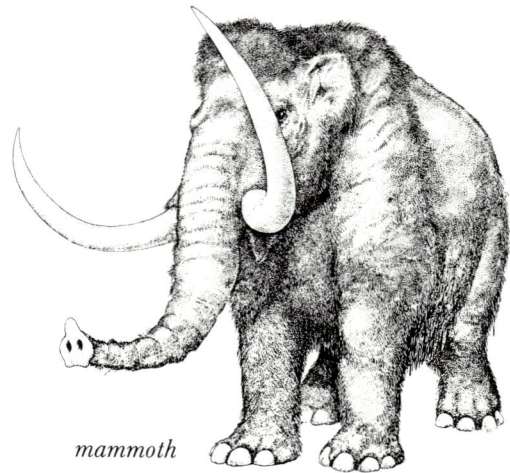
mammoth

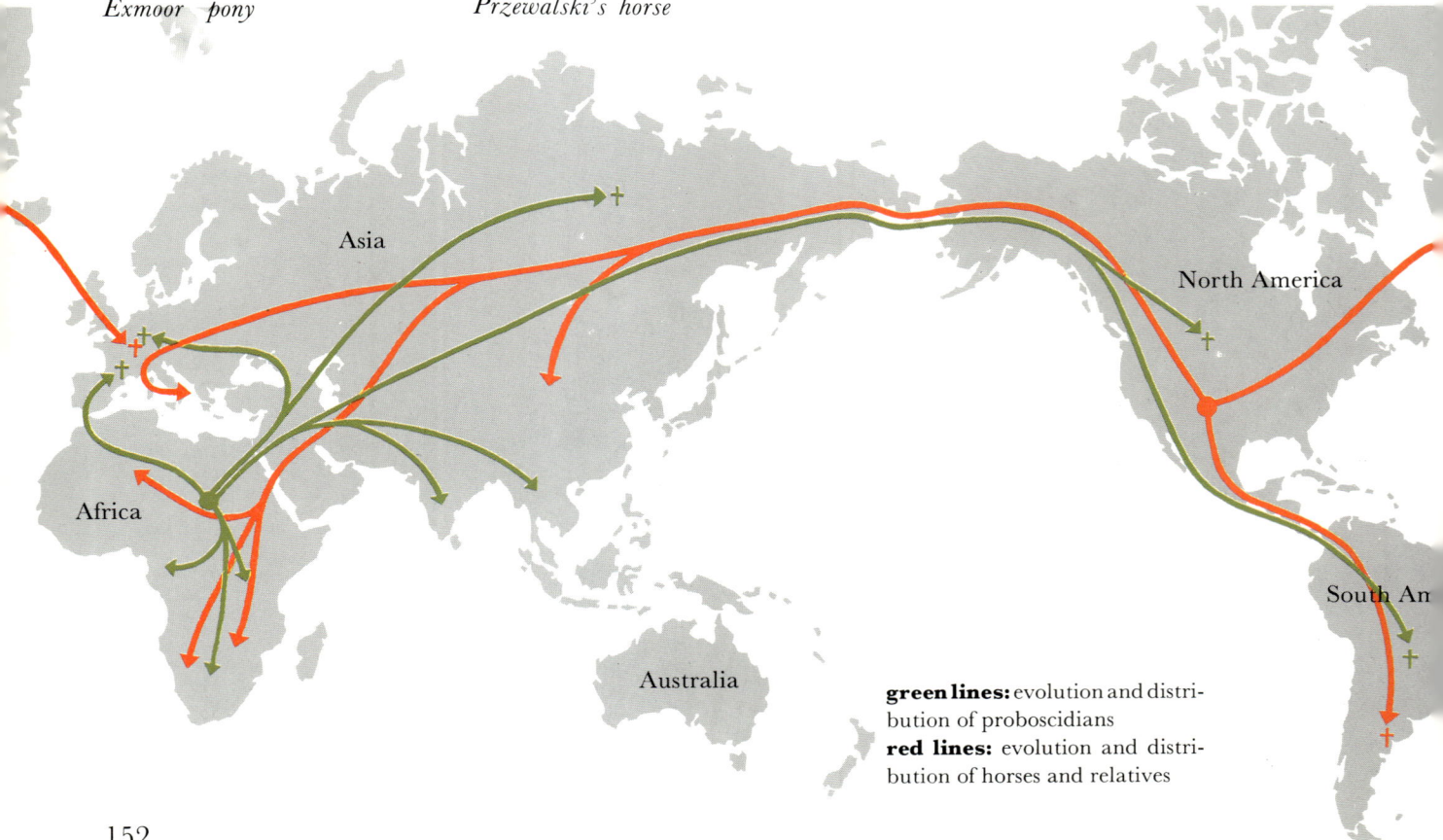

green lines: evolution and distribution of proboscidians
red lines: evolution and distribution of horses and relatives

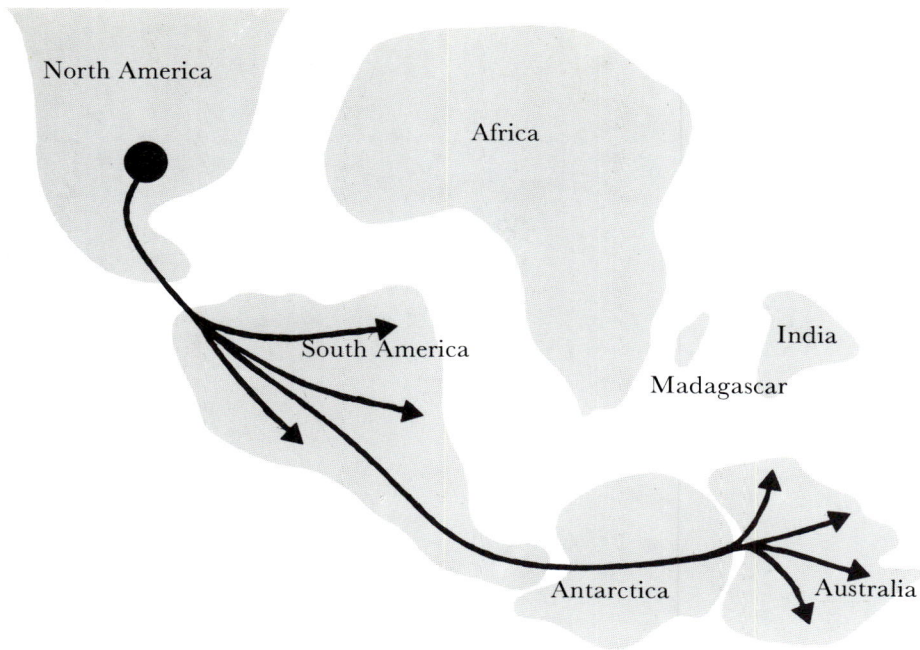

North America

Africa

South America

India

Madagascar

Antarctica

Australia

The marsupials existed as a group approximately 180 million years ago in North America. In the following 30 million years they spread over the then existing land bridge to South America and over the still ice-free Antarctica to Australia (map, left). About 65 million years ago, Australia became separated from Antarctica.

Now that Alfred Wegener's continental drift theory has generally been accepted, we know that the earth's development was far more dramatic than was previously suspected. One huge primitive continent divided into two parts, henceforth separated by oceans. Changes in the thickness of the ice cover of certain regions led to rises and falls in sea levels, causing isthmuses and straits to

of Antarctica, then free of ice, for more than thirty million years. It was by way of this land bridge that the marsupials reached Australia (map, above). Then, throughout the Tertiary, South America remained isolated. The isthmus joining it with North America dates back only one and a half million years. Only then were modern mammals able to invade South America.

opossum

Tasmanian devil

kangaroo

koala

appear. And these geological upheavals naturally affected the expansion of animals.

We now possess accurate information about the prehistoric migrations of land animals because of the discovery of numerous fossil deposits. Furthermore, we can see, much more clearly than we can for the birds, how the land animals extended their range, since for them the sea was an impassable obstacle. After being separated from Africa several hundreds of millions of years ago, it remained linked to Australia by the continent

As South America was isolated in the Tertiary, its marsupial fauna evolved as freely as in Australia. Many of these species were, however, supplanted by modern forms before the emergence of a land bridge to North America about one and a half million years ago. But the opossum persisted and colonized North America.

The small Tasmanian devil is one of the few predatory marsupials to survive today.

Until Europeans arrived in Australia, the kangaroos, as grazers, played the same role as hoofed animals elsewhere.

The koala is a specialized leaf-eater, adapted in habits and digestive system to the Australian eucalyptus tree.

153

AT THE LIMITS
OF LIFE

Approximately forty per cent of the earth's land surface consists of desert – barren, desolate regions with extreme climatic conditions. Merciless heat and, even more intolerable, severe drought in the hot desert areas, murderous cold and unimaginable storms in the cold deserts of the polar regions and in high mountains make it

that animals and plants have occupied every ecological niche.

Micro-organisms have been particularly successful in overcoming these extreme conditions. In the excessively cold and dry icy wastes of Antarctica, where the annual mean temperature is about $-55°C$ and there is no life on the surface, blue-green algae and tiny

Left: in sandy deserts, apparently completely dead, a downpour of rain can work wonders: in a very short time plants appear from the bare ground. They come from seeds which have perhaps remained embedded in the soil for years, and they serve as food for insects which have also waited for the rains in order to hatch from eggs.

Right: there is even life in the icy polar wastes. Apart from a few micro-organisms, only warm-blooded animals are capable of withstanding the very low temperatures, largely thanks to their ability to maintain their body temperature.

extremely difficult for animals and plants to survive. In many deserts there may be no rainfall for several years and then a brief deluge. In such zones survival is only possible by virtue of an assortment of amazingly sophisticated adaptations. It is even more astonishing that so many living organisms have been able to overcome the natural difficulties of such environments, so

lichens have settled inside rocks from which they dissolve all the sustenance that they need. Nor is the Dead Sea literally dead. Here, about 400m below sea level, in water that is four times as salty as that of the ocean, harbouring no fish, live several species of bacterium and the alga *Dunaliella*. Bacteria can even survive in the boiling water of hot springs.

155

SIGNS OF LIFE

There are certain desert areas, either of sand or ice, in which a search for signs of life would reveal, at most, a few microscopically small organisms. But such 'dead' wildernesses are rare. Most deserts harbour a surprisingly large number of higher organisms as well. But even in areas where there is a comparative abundance of plant and animal life, a casual observer will see little except sand and rocks. Most of the plants are inconspicuous. Some look like the stones among which they grow. Others only appear after rain has fallen and then they flower for the briefest period, their seeds lying dormant in the soil, for months or years, until rain falls again and they germinate. Many insects have the same life cycle. Their eggs may, if necessary, remain viable for years. The larvae only hatch when the soil contains sufficient water to guarantee adequate vegetation on which they can feed. Many animals survive dry periods in a state of suspension with greatly reduced metabolism. However, there are also species which remain active throughout the year, but live a concealed life. Most of them hide away by day in rock crevices or in the sand to avoid the heat. They come out at night to search for food and disappear again in the morning. Only tracks in the sand betray their presence.

The polar regions, too, often appear dead. The land animals that live here, predators and prey, cannot be conspicuous and many of them are white.

Polar regions (above) and hot sandy deserts (right) usually appear to be devoid of life, but in the endless snow fields and in the flat, uniform sandy wastes there are tracks to indicate that even here in the wilderness animals live a secret, hidden existence.

156

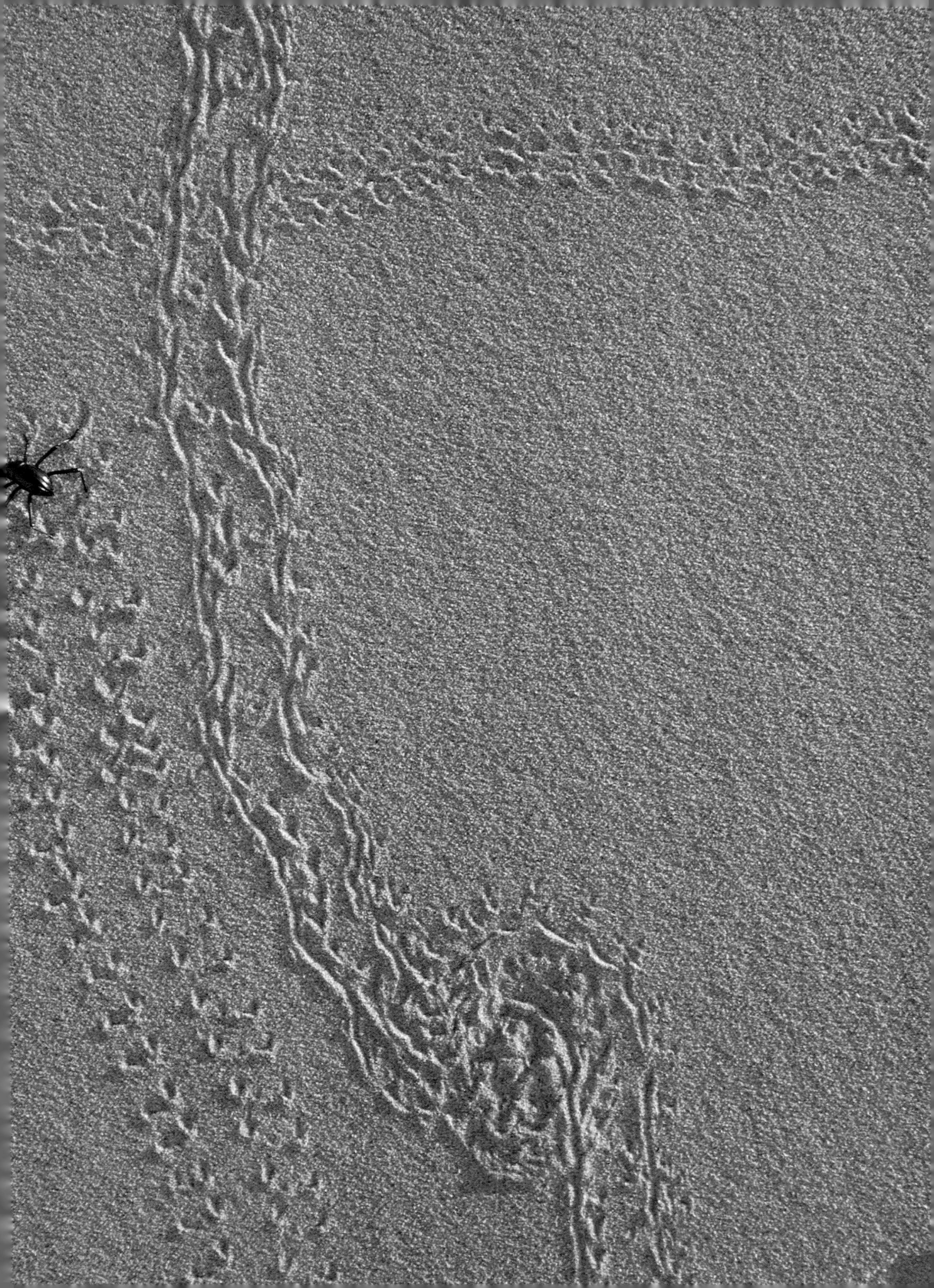

THE BOUNDS OF THE POSSIBLE

Above, right: the Rocky Mountain goat almost always lives above the tree limit in North America. In the Rocky Mountain region it survives the winter, even when practically all other animals have retreated to the valley or have hibernated.

Above: even in the spray zone of steep cliffs, subjected to the pounding of the waves, there are numerous animal species, such as barnacles, looking rather like molluscs, but actually crustaceans, and limpets which can adhere very firmly to the rocks.

The numerous animal species which occur in extreme climatic zones show a great range of adaptive strategies to enable them to survive.

The four seal species living in Antarctica have different hunting methods. The large and slender leopard seal, a solitary animal, is so fast that it can catch penguins in the water, though it feeds more frequently on fish and invertebrates. The rarely observed Ross seal, which lives on the pack ice, has strikingly large eyes which enable it to catch fish in the dim light under the ice. The crab-eater seal, despite its name, only consumes the tiny crustaceans known as krill, retaining this food by means of cusps on its teeth and sieving out the water. Finally, the Weddell seal feeds on bottom-living fishes and squids, which it sometimes brings up from depths of over 500 metres. While the other seals can dive for about a quarter of an hour, the Weddell seal can, if necessary, remain underwater for up to an hour. This singular capacity, for a lung-breathing mammal, is explained by the fact that it has blood containing about five times as much oxygen as human blood. Furthermore, in the course of its dive, only those organs that would quickly die without oxygen, such as the brain, are supplied with blood. The Weddell seal hunts its prey in the darkness as bats do in caves, giving out squeaks and locating the fishes by reflected echo.

Astonishingly few polar animals hibernate. Rodents, such as lemmings, dig burrows in the surface soil where, protected by a thick blanket of snow, they find enough plant food to survive. Their enemy, the arctic fox, also remains active throughout the year. It digs itself into the snow to sleep, even though it can withstand bitter temperatures of about 50°C below zero.

A number of plant-eating insects species occur in desert zones where, contradictory as it may seem, no plants grow. They can remain active in areas in which literally no vegetation is visible for months or even years on account of the drought, by feeding on tiny particles of dead plants which the wind has transported over long distances from more fertile areas and deposited in the lee of sand dunes. Most of the insects that feed on these scarcely visible plant particles are beetles. In the Namib desert of south-west Africa, one of the driest areas in the world, where it may rain only once in ten years, no fewer than two hundred of these resourceful beetle species have been identified. Their method of solving the food problem is quite extraordinary. When a westerly wind from the Atlantic sweeps drifts of cloud over the desert, the beetles stand on the ridges of the sand dunes, sometimes forming long rows with their heads to the wind and their abdomens raised on the long rear legs. The mist settles on the backs of the beetles where it forms small droplets which run down into the mouth.

Left: snakes even occur in rocky areas and desolate lava fields. They consume insect-feeding lizards or small rodents which can live in the most varied habitats thanks to their astonishing ability to adapt.

Below: musk-oxen, only 1.3m tall, have a thick pelt with hairs up to 90cm long, giving them a more massive appearance. These animals can survive even the coldest arctic winter.

159

BEYOND THE POLAR CIRCLES

A dense pelt and a thick layer of fat protect the polar bear from the arctic cold. It can even tolerate long periods in the water. Although really a land predator, it sometimes seizes seals from the water close to the shore, swimming under the surface and coming up just in front of its victim. It can leap out of the water to a height of 2.5m.

Like all desert animals, the creatures of the polar regions also have to be adapted for extreme environmental conditions. Under these conditions life is no longer possible for cold-blooded vertebrates in which the body temperature is always more or less the same as the outside temperature. The adder is one of the very few species whose range extends just beyond the Arctic Circle.

Nevertheless, it is only active for a few months in the year. In places where plants can still grow, there will also be insects, particularly springtails and mosquitoes. Their development is completed within a few weeks, so a short period of warmth is sufficient for survival.

Warm-blooded animals, birds and mammals, can maintain their body temperature around 37°C, even when the external temperature is lower. To avoid too great an energy loss, polar animals have thick plumage or fur capable of retaining a quantity of air, and often a dense insulating layer of fat beneath the skin. They also have a small surface area in proportion to body volume because they are more stoutly built and usually larger than their counterparts in warmer regions. In order to avoid unnecessary heat loss by radiation, the extremities and the ear lobes are as small as possible and roundish. Naked areas of skin such as the soles of the feet, the flippers and the nostrils are kept cool by a special system, which ensures that they lose as little body heat as possible; the arteries and veins leading to these parts are intermingled to form a kind of sphere. As a consequence, warm blood coming from the heart heats the cold blood of the extremities, while the latter simultaneously cools the warm arterial blood.

Most polar birds are fishers. A few species catch their food either at the surface or by diving. Others, such as penguins and the very similar looking auks, are true underwater hunters. In the course of their evolution, penguins have lost the power of flight and their wings have become particularly efficient swimming flippers. They do not need the power of flight in the Antarctic where there are no land predators.

Left: emperor penguins breed in the antarctic winter. After two months' incubation and five months' rearing, the young are finally able to swim and dive, before they become independent in summer.

Below: in contrast to most of their relatives, snowy owls hunt by day, because the small animals on which they feed are also active in the daytime.

Above: the arctic fox, exhibiting a white and a dark-coloured phase, is extraordinarily adaptable. It can even live on the drift-ice, feeding on all kinds of animal remains, including droppings.

ROCK, SNOW AND ICE

The atmosphere surrounds the earth like a protective sheath. It ensures that the earth's surface does not cool off too rapidly at night and it holds back the potentially injurious part of the sun's rays.

The highest mountains rise into the upper layers of the atmosphere. Any organism living in these zones is faced with extreme conditions similar to those endured by desert and polar animals. There are very harsh temperatures and frequent changes of weather. At heights of 4000 metres in the

Below: with its very dense fur and broad feet which stop it from sinking into the snow, the snow leopard is well adapted for its home range, the mountains of central Asia. In summer it lives at 3000–6000m, in winter at about 2000m.

red blood corpuscles to take up the oxygen: the vicuña, living in the Andes, often at heights above 5000 metres, has about three times as many as a human being – about fifteen million per cubic millimetre.

Cold blooded animals find life here as difficult as in the polar regions, but there are insects up near the snow-line, including a flea that feeds on wind-borne pollen.

Right: large-toothed hyraxes usually live at low altitudes, although their colonies can be found at about 4000m on Mount Kenya. Here they have a chance of avoiding leopards, their main enemy.

tropics the sun may make it very warm during the day while at the same time the snow never melts in shaded places and the night temperature may fall to more than 10° below freezing point. At this height, too, there is a scarcity of oxygen. Therefore, most high mountain animals possess particularly efficient heart and lungs, have a larger quantity of blood than lowland animals of the same size – and, most importantly, more

Even at lower altitudes, the development of eggs without body heat is no longer possible, so mountain reptiles such as the adder and viviparous lizard carry the eggs in their bodies and give birth to live young. The alpine salamander, an amphibian, whose young live in water as gill-breathing larvae, does not give birth until the end of the larval stage. By contrast, the Andean toad can live amid ice and snow, breeding in and around hot mountain springs.

Above: in autumn ptarmigans exchange their grey-brown plumage for white. In winter they dig long, deep tunnels in the snow and are thus still able to find edible plant shoots and to survive at temperatures of minus 40°C.

Far left: the toes of chamois are adapted for steep country. The two main claws on each foot are movable, the soles elastic, the edges hard and sharp; and the tips of the smaller hind claws dig into the ground.

Left: ibex are even more expert climbers than the chamois. Whereas chamois normally live on slopes just above the tree limit, ibex are often found in the most treacherous rocky terrain: in summer, in the Alps, up to an altitude of about 3500m.

163

LIFE-GIVING WATER

Above: the apparently desiccated leaves of bushes in dry semi-deserts are sufficient to satisfy the fluid requirements of the gerenuk, which has never been seen drinking water. Thanks to the long neck and the ability to stand on its hind legs, it can browse on twigs 2.5m above the ground.

There are several ways in which an animal can keep its body cool in intense heat – by hanging out its tongue and panting, by licking the parts it can reach, if necessary with the aid of the paws, and by sweating. The last is especially effective, but mammals in hot deserts are denied this luxury because they have no sweat glands. Some, like the Sahara fennec and the American jack rabbit, have huge ears with blood vessels close to the surface, so that the blood is cooled by any breath of air. But the majority, able to tolerate more heat than animals in temperate climes, remain hidden by day in burrows where the temperature rarely exceeds 35°C. Rodents, of which there are hundreds of species in the desert, block the entrances to their holes, often a metre deep, to keep out the heat and retain humidity.

Finding water, and conserving energy while doing so, calls for even more ingenious solutions. For flesh-eaters such as jackals and fennecs, it is relatively easy because their food contains a lot of fluid. Radical adaptations are necessary, however, for those animals which live mainly on plants poor in water or even exclusively on seeds, which contain only about 10 per cent water.

The important thing for a desert animal is to conserve as much liquid in the body as possible. In some the specialized nostril mechanism removes the humidity from exhaled air and returns this to the body. And many rodents, like the

American kangaroo rat and its African and Asian counterparts, the various gerbils and jerboas, similarly built with highly developed hind legs and long, hairy tail, can do without drinking water altogether, using up the tiniest amounts of fluid contained in the seeds they eat. They also produce urine which is some five times more concentrated than that of humans, retaining fluid thanks to the filtering capacity of the kidneys. Practically every drop of

Left: desert geckos can survive in a resting state for long periods, thanks to the fact that they accumulate a large fat reserve in the tail.

Above: male sand-grouse drinking in shallow water. Facing page, below: an adult sand-grouse with chick.

fluid in the faeces is also extracted in the large intestine, and often the hard droppings are also reswallowed by these animals.

For animals such as amphibians, which are very dependent upon water, the desert is naturally a poor living environment. Nevertheless, there are quite a few species, particularly toads, which survive in deserts. By day they remain deep down in their burrows. During periods of drought many species aestivate, a condition comparable with the hibernation of temperate zone animals. Some can lay down a store of water in the body. For spawning they usually do not use the few water-filled lakes because most of these are salty. When rain forms pools, the sex drive of these amphibians is immediately stimulated. Large groups come to the water, mate and spawn. Quite soon, after only

two days, the larvae hatch and start developing. They burrow into the bottom and feed on organic matter, mainly algae, and grow rapidly. In certain species, when the water dries up too soon, some of the larvae grow modified jaws, become cannibalistic and eat their defenceless companions. This increases their chance of reaching the end of the larval stage. Then, almost at the last moment, they become full-grown toads, leaving the mud, now nearly dry, and burrowing in the sand.

Left: elephant shrews do not have to drink water. By a chemical process they can convert some of their body fat into liquid.

SURVIVAL IN THE DESERT

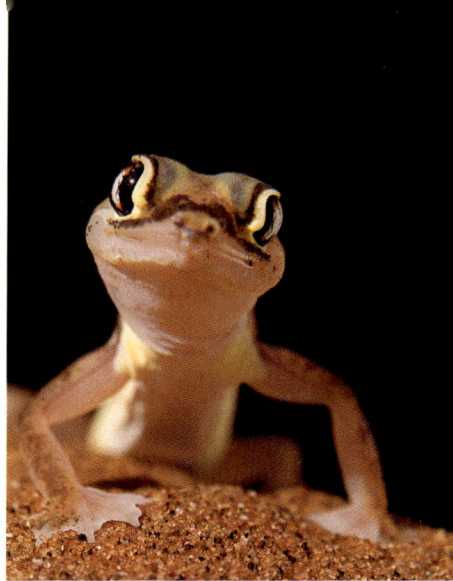

Right: the desert gecko of Namibia is nocturnal. It has webs between the toes, which are used like snow-shoes to prevent it sinking into the sand and which serve as shovels for digging.

This grasshopper has an adaptation very similar to that of the gecko for living in sandy deserts. With its enlarged feet, it walks about on sand, mainly at night.

In some deserts the heat of the surface sand by day may rise to over 70°C. Beetles and grasshoppers that live there often have long legs like stilts, with which they can raise their bodies off the burning sand. Some of them also have remarkable tiny branch-like structures on the tips of their feet, which function like snow-shoes and prevent them sinking into the sand. Lizards which are active by day also move around as though on stilts. At intervals they lie flat on the belly for a few moments, raising their legs and tail to cool them a little.

The majority of desert animals, however, are nocturnal and disappear in the morning twilight. Various adaptations of limbs and body enable them to sink effortlessly into the sand and spend the day in a more tolerable environment, because at 30 centimetres or so below the surface the temperature seldom rises above 30°C. Desert scorpions use their particularly large pincers to dig down to a depth of about 75 centimetres where it is even cooler. Thanks to their flat, lens-shaped body, certain beetles can vanish rapidly into the sand with a couple of sideways movements. Geckos have broad, fringed or even webbed toes on all

four feet which serve as snow-shoes on the surface of the sand and as flippers, comparable to the fins of fishes, when they burrow. Most lizards, however, get about, with wriggling movements of the body, on the same principle as a fish in

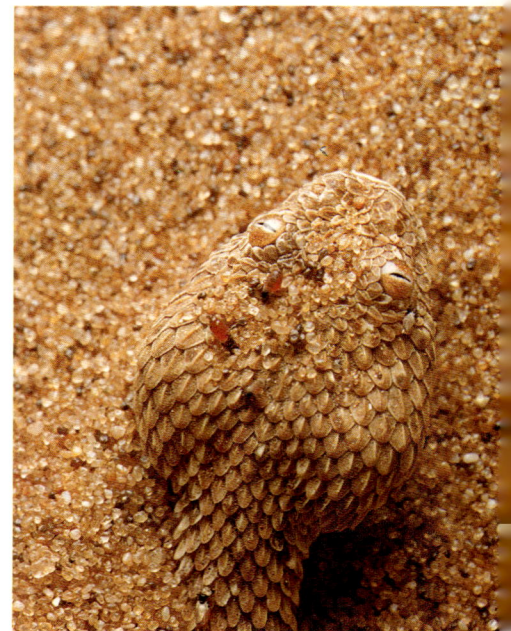

water. The skinks, for example, are typical 'sand-swimmers'. In order to have as little resistance as possible in the sand, these lizards have scales with a completely smooth upper surface, and tiny legs which they hold against the body, enabling them to move swiftly and efficiently through the sand. Indeed, one skink species in the Sahara has been dubbed 'sand fish'. In the Namib, another skink that lives permanently under the surface has lost its legs altogether.

Snakes, too, use sideways body movements to travel rapidly over soft sand, leaving clear tracks.

Two snake species, not closely related, the South African dwarf puff adder (far left) and the North American sidewinder (left) have evolved the same method of moving, alternately lifting and lowering the front and rear parts of the body. Lifting the body also helps to cool it a little. By means of these sideways movements both species can sink into the sand in less than a minute.

Below: the fringe-fingered lizard has toes widened by comb-like scales. When walking, it lifts its feet very slowly, thus cooling them. It may also lie for a few moments on its belly, with all four feet in the air to cool. When too hot, it buries itself in the sand.

SURVIVAL IN MOUNTAIN STREAMS

The larva of the Egdyonurus *mayfly (top) has a much flattened body, enabling it to survive in fast-flowing waters. The flat stonefly larva (above) has two claws on each foot, with which it fastens itself to the bottom. In the larva of the dryopid beetle (right) the whole underside forms a suction disc. The larvae of net-winged midges of the* Blepharoceridae *(far right) have six small suckers on the underside. The pupae are attached to stones by a glue-like secretion.*

Mountain streams are extremely difficult places to colonize and it is a wonder that animal species have managed to do so and thrive. Not only is the water always very cold but, more to the point, it flows terribly fast. Higher water plants have no chance of survival, and only the tiniest algae can develop by clinging to the stony stream bed. Moreover, only a few fish species, such as salmon or trout, are strong enough to contend with

these currents and to swim about freely.

Practically all other animal species frequenting these zones have special adaptations which enable them to attach themselves to rocks and stones. There are certain very broad, flattened fish species in which the entire underside acts as a suction pad, and these move by literally creeping over the rocks. Some insect larvae, such as those of mayflies and dryopid

beetles, do the same. More surprisingly, the larvae of certain frog species live in these rough waters. In one species the mouth faces downwards, and while they browse the algal mat with their tiny teeth they hold on with the enormous fleshy lips, which act as a suction cup. Even better adapted are the tadpoles of the *Staurois* frogs of South-east Asia; behind the mouth they have a suction disc which covers most of the belly.

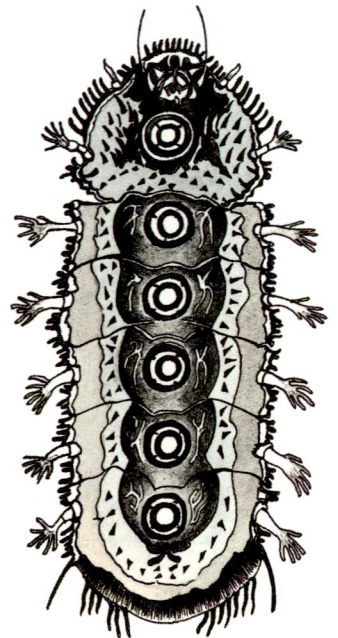

The dipper is an astonishing bird which lives in fast-flowing streams. Searching for insect larvae, it runs along the river bed, gripping the rocks with its toes and, if necessary, paddling with its wings. Even more extraordinary is the torrent duck of the Andes, with an exceptionally long and streamlined body, which swims, entirely submerged except for its head, moving with amazing speed, even against the current.

LIFE IN THE SPRAY ZONE

Below: mussels, which attach themselves to rocks by special byssus threads, live in sheltered crevices. Barnacles (right) can also withstand the breakers, as their chalky shell moulds itself perfectly to the rocks. The animals which form these shells are actually crustaceans.

There are three different types of seashore: sandy, muddy and rocky. Each represents a specific biotope. In the intertidal zone of the first and second type, conditions are challenging enough for the first two types because animals living there have to be amphibious and acquire other adaptations as well. In the spray zone on a rocky coast conditions are even more difficult and, for most species, insurmountable. Nevertheless, there are certain animals which resolutely survive here, defying the waves which regularly thunder against the rocks and, by special adaptations, protecting themselves against the scorching sun; for drought constitutes an even greater peril than the continuous battering of the waves.

Clefts in the rocks are favourite shelters for intertidal animals because they offer protection against both the breakers and the drought. Here mussels, sea-stars and sea-urchins, sea-anemones, snails and various crustaceans can live, sometimes in very large numbers. Many encapsulate themselves in hard, close-fitting armour, and thus avoid desiccation when the tide goes out. Limpets are particularly well adapted and do not have to seek shelter. Their flat, cup-shaped shell fastens them firmly to exposed surfaces in absolute safety. At low tide limpets creep about, browsing on algae, and before the tide comes in again they have returned to their home site. Here the constant rubbing of the shell on the rock excavates a shallow depression which fits the body perfectly. In this way the limpets keep themselves moist and at the same time remain tightly attached. Other animals attach themselves differently. Acorn barnacles secrete a sticky substance in their carapace which hardens against the rock; mussels cling to the surface with a special organ, the byssus.

Right: eight hooked feet and a flattened body prevent this crab from being swept away by the waves.

On rocky coasts the intertidal zone offers a wealth of food. Algae grow over the rocks and each wave carries innumerable new, protein-rich organisms. Animals wanting to utilize this inexhaustible food source have to be highly adaptable; they must not only master the problems of amphibious life, but also be able to withstand the crashing waves.

HIBERNATION

Below: in temperate zones, during autumn, bats seek retreats, often far from their summer quarters, usually in caves but also in hollow trees or cellars. Here they fall into a lethargic state of hibernation with minimum metabolism.

Birds do not need to hibernate. Their ability to fly and their sense of direction enable them to move off at the approach of winter to regions that are climatically more favourable. There is, however, one bird species which stays put and hibernates in the true sense; this is the poorwill, a North American nightjar, which finds a frost-free retreat and hibernates there for about three months, reducing its bodily functions to a minimum, and making use of fat reserves to provide the energy necessary for its survival.

It is surprising that only a few insectivorous bats in temperate latitudes migrate. Most of them hibernate in frost-free caves although, like the birds, they would have no difficulty in flying off to warmer climes.

The only true hibernators are certain warm-blooded mammals; and even in their case the 'depth' of the winter sleep may vary considerably. Brown bears, for instance, are easily woken and often interrupt their sleep. Thus, they keep their body temperature high, and their pulse rate, normally forty beats per minute during sleep, does not fall below ten beats per minute. In most hibernating mammals the body temperature sinks considerably, but it then remains constant and is not dependent upon the outside temperature. At high altitudes a marmot may hibernate for over six months with a winter body temperature of only 5–8°C. Its pulse sinks from an average of 115 to only four beats per minute, the oxygen demand is reduced to about one-twentieth of normal requirements and it takes only two or three breaths per minute. Mammals in this condition are cold and stiff to the touch, to all appearances dead. A hibernating dormouse can be rolled along the ground like a ball, and will not wake up. Most hibernating warm-blooded animals wake as soon as there is a marked

Centre: the dormouse retreats into hibernation at the end of September. It is only half as heavy when it wakes 6–7 months later.

Far right: hedgehogs wake up, not only when the weather becomes warmer, but also when their body temperature falls below 5°C. If it can reach the correct temperature again, by increased metabolism, the animal goes back to sleep. This prevents it from freezing.

Left: marmots sleep in densely packed groups for 6–8 months in sealed burrows. About once a month they wake up to defecate and urinate, but without feeding. They live on their fat reserves which in autumn may account for about half their body weight.

Above: brown bears spend most of the winter in hiding. This is not true hibernation with a much reduced metabolism. They maintain the normal body temperature and will wake up at the slightest disturbance.

rise in temperature, but also if the temperature unexpectedly sinks lower than normal. If they did not react to this stimulus, they would freeze while asleep in extreme cold. By moving about, they can warm up and, if necessary, seek a warmer retreat.

The phenomenon of hibernation in mammals involves complicated, hormonally controlled, chemical processes. While it is in progress, for example, the blood contains about half as much sugar but twice as much magnesium as in summer; furthermore, it is far less apt to clot, for otherwise the slowing down of the blood circulation might cause thrombosis.

The winter sleep process is a much simpler matter for cold-blooded animals such as amphibians, reptiles and insects. Their temperature and body functions are reduced along with the outside temperature, unchanged by any internal activity. Few can survive sub-zero temperatures.

173

FOOD STORES

Far right, above: camels can survive in extremely dry areas. For this they store large amounts of fat in their humps.

Right: among honeypot ants, certain workers hang from the roof of the nest, where they are fed with honey until their abdomen becomes swollen to a sphere. They provide food for the colony when flowers are scarce.

Far right, below: autumn marmots carry large amounts of grass into their burrow. This is not, as was long thought, used as food, but serves to upholster the nest and to close the burrow entrance. Marmots live throughout the winter solely on their fat reserves.

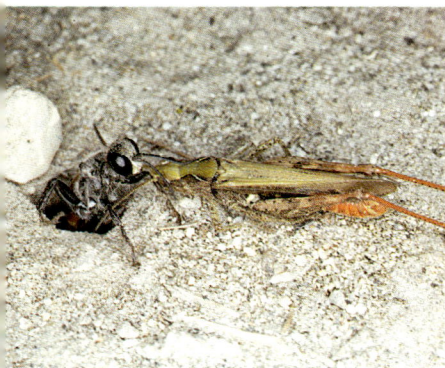

Above: a sand wasp drags a grasshopper it has paralyzed into its previously prepared brood chamber, lays an egg on its victim, leaves the hole and closes it with a suitable stone. The wasp larva feeds on the grasshopper.

Above, centre: a non-social bee returns to its nest with a load of pollen.

The true hibernators among the mammals which sleep for several months without waking develop a thick layer of fat during the autumn which serves them as an energy store until the spring. It is this indispensable quantity of fat which, by activating certain glands, actually determines the onset of hibernation, whereas in cold-blooded animals the timing depends on an appreciable drop in outside temperature.

Many animals interrupt hibernation quite frequently. Some do not sleep more than they normally would, but lay in stores so that they do not starve. Among the insects the best-known example of store-keeping is by honey bees but many other insect species make provision in a similar way.

Among mammals, hamsters are probably the most assiduous food collectors. The common hamster, weighing about 300 grams, carries food in its inflatable cheek pouches and is capable of bringing 50 kilograms or more of grain and other wild seeds into its burrow. Many other rodents have similar cheek pouches that enable them to store sufficient provisions for their winter needs. Mice and rats will hoard as ravenously as squirrels, though the latter do their nut and acorn collecting more conspicuously.

The nutcracker has a collecting pouch in the form of a throat sac, capable of holding several dozen seeds of the stone pine tree. In autumn it buries these at various places in its territory. When winter comes, it retrieves most of these seeds with amazing precision, even when they are under thick snow. Obviously, it cannot find them all, so the bird helps indirectly to distribute the stone pine.

Martens, foxes, wolves, polar bears, arctic foxes, cats, owls and other animals also lay down stores

Left: the American ant-eating woodpecker stores vast quantities of acorns for the winter. It chisels holes in the bark of a tree and inserts the nuts in them.

Above: the Australian butcher-bird catches various small vertebrates, mainly birds. It makes a food store by impaling the victims on thorns or in a forked branch.

of meat when the hunting is good, but as a rule these reserves scarcely last for more than a week. Polecats are said to paralyze frogs by a bite in the neck and to distribute up to a hundred victims in various hiding places, but this has not been verified. Moles, however, are known to do exactly the same to earthworms, sometimes collecting a thousand or more in clumps the size of a fist. This is not an autumn activity in preparation for winter; in fact, they pile up stocks mainly during winter when, like their prey, they live in the frost-free layers of the earth where there are plenty of earthworms.

Below: the crop of the nutcracker is spacious enough to contain more than 100 stone pine seeds. It buries these and recovers most of them later on, even when there is thick snow cover. The remaining seeds help to disseminate the stone pine.

TECHNICAL FEATS

Television transmitters, jet engines and materials reinforced with glass fibres are, in their very different ways, masterpieces of technology. But are these genuinely original human inventions? The truth is that many devices and processes allegedly discovered by man were already present in the natural world. And we can still learn a good deal from the study of animal and plant structure. Blades of corn, for example, are built on the same principle as television transmitters, although they are proportionally lighter and more slender. Dragonfly nymphs and squids have long since employed jet propulsion for their normal movements. The principle of fibre reinforcement has been anticipated in the structure of plant cell walls. Indeed, many millions of years of life and death cycles in the natural world have provided the solution to virtually every problem that we ourselves are faced with and are belatedly trying to tackle.

Living organisms have shown particular advances in the realm of chemistry. With the aid of their enzymes, animal and plant cells have managed to achieve many chemical reactions at room temperature. For this human technology requires great heat or high pressure. The modern use of micro-organisms for the industrial production of materials is a case of exploiting Nature's know-how.

Left: like all bats of the family Vespertilionidae, the long-eared bat (Plecotus auritus) emits its direction-finding ultra-sounds through the open mouth. The nose is not involved in sound production, as it is in horseshoe bats. The weak echoes are picked up by the ears. The function of the flaps in front of the ear-hole is not yet clear.

Right: this radiolarian skeleton came from a deep-sea core taken by the research vessel Glomar Challenger *off Timor. Like their present-day descendants, these animals, dating back millions of years, floated freely in the ocean. Their skeleton anticipates the advances of modern constructional engineering — lightness in combination with strength.*

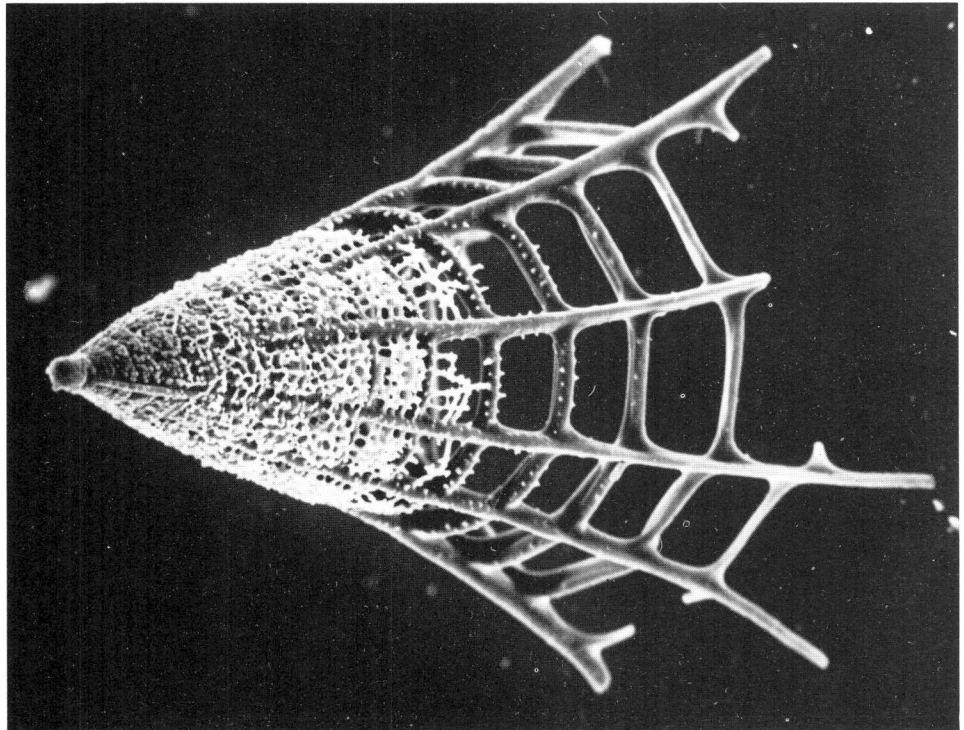

CYBERNETICS OF LIFE

In homoiothermal ("warmblooded") animals the maintenance of a constant body temperature is a classic biological control process. Control is exercised on two planes, behavioural and physiological. Depending upon the outside temperature, animals seek out a sheltered or exposed, shady or sunny place. Meanwhile, a centre in the brain constantly measures the blood temperature. If this is too low, the body is warmed up by shivering and increased metabolic activity. Animals keep cool in different

ways. Horses, for example, lower their temperature as we do, by sweating, whereas dogs can only cool off by panting.

The term cybernetics comes from the Greek word for pilot or helmsman. The helmsman is responsible for making sure that his ship follows a defined course. He repeatedly measures the position, e.g. with a sextant, and compares the actual course with the planned route which he reads from a sea chart. If the ship diverges from this course, he uses the rudder to bring it back to the correct path. Itineraries may have become more complex, instruments more sophisticated, but the ancient principle still applies.

What has all this to do with cybernetics? To put it very simply, cybernetics is the science of control systems, and the case described above is a classic control process or model. Such models consist of a few basic elements: a governor (here the helmsman) receives from a sensor (the sextant) information on the actual course. He tries to bring this actual course into agreement with the desired course plotted on the sea chart. To do this he uses the rudder of the ship. Control systems are widely used in model. Such models consist of a example, a thermostat is a governor or regulator. Using a thermometer as a sensor, it measures the room temperature and by switching the heating on and off at suitable intervals, maintains the temperature as close as possible to the desired value. The result is precisely that achieved by the ship's rudder.

The two examples described above from the field of technology – the control of a ship's direction and of a room's temperature – show, in extremely simplified form, how variable these systems may be. Thanks to its universality, cybernetics serves as a link between the most diverse sciences. And in living organisms, similar control systems, in their own way equally sophisticated, can likewise be identified and analysed.

In fact, such control systems are a fundamental prerequisite for all forms of life. Humans and all other living organisms would be incapable of existing even for a moment if it were not for a great number of control systems which automatically and continuously regulate changes within the body, reconciling actual and ideal values. For example, the human body temperature is normally kept at a desired value of 36.9°C. Fluctuations of one degree above or below may lead to discomfort, and greater fluctuations to illness, even death. In this case, the regulator is in the middle brain where there is a sensor which continuously measures the temperature of the circulating blood. At the same time the regulator receives information about skin temperature from the cold receptors in the skin. If the blood temperature exceeds an upper limit value of 37.15°C, we start to sweat and this has the effect of cooling us down. On the other hand, if the skin's cold receptors report a disturbing or dangerous drop in surface temperature, then the regulator stimulates the control of the metabolism,

Left: honey-bees regulate the temperature in the hive within about one degree. To cool off, they release water drops from the mouth and ventilate by wing beats. To warm up, they vibrate the thoracic muscles.

and makes the muscles work: the body shivers from the cold and we warm up. This brings the actual, deviating body temperature back to normal, 36.9°C.

Other bodily functions are subject to similar controls as, for example, the oxygen supply that we need for breathing, the level of sugar in our blood and the nerve centres that enable us to keep our balance when we are standing. In each of the cells of our body there are thousands of chemical control processes at work every second.

Control systems also operate in the area of animal and human behaviour. We can consciously influence and thereby regulate our body temperature by lying in the sun to warm up or retreating to the shade when we become too hot. In such a simple action as reaching for a glass of water we make use both of our eyes and our sense of touch – the sensor that relays the information.

Cybernetics also has a role to play in the wider field of environmental conservation. For all that man has done to interfere, a tenuous equilibrium has been maintained. The question is – will it be possible to devise a control system that will ensure the very survival of animal and human life?

Above: flamingos can stand for hours at a time on the very hot surface of salt lakes, yet they are also found in mountain areas on ice and snow. To withstand such contrasts they have a special network of blood vessels in their legs, which prevents the feet getting too hot or too cold.

Control systems are not confined to the body. The very environment is self-regulating, and every living organism has its function. Even predators are indispensable for keeping the populations of their prey healthy, by killing the weak and sick animals.

WINGS

Right: wing of an African white-backed vulture. The Old World vultures are poor fliers, tiring quickly and hardly able to take off after gorging themselves. For hours at a time they circle around, using thermal currents to raise their large wings.

Vertebrate animals have, at different times, developed three forms of flight. In bats the flight membrane is stretched between the second, third, fourth and fifth fingers, leaving the first finger free (above). In birds (centre) the wing feathers are

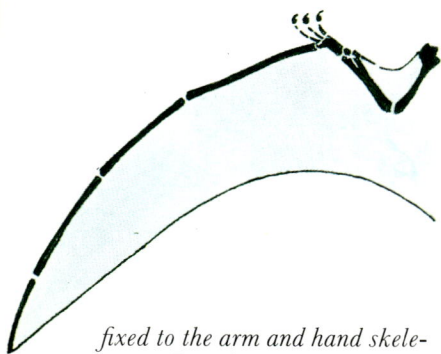

fixed to the arm and hand skeleton. In the extinct flying saurians (below), the much enlarged fourth finger alone supported the whole flight membrane; the first three fingers were normally developed and probably used for climbing.

Some 400 million years ago, the first animals to take to the air were insects. This was a considerable step forward, since for the same expenditure of energy, a winged creature could now cover a much greater distance than a comparably sized land animal. Insects have an external skeleton, and their wings are simply skin structures, extensions of the thorax.

Vertebrate animals took somewhat longer to fly and actually succeeded in doing so on three occasions. The first, about 200 million years ago, was when saurian reptiles took to the air. Their flight membrane was stretched between the much elongated fourth finger and the body, the other three fingers being normally developed. There were several species of flying reptiles, from sparrow size to those with a wingspan of over twelve metres. Birds, which first appeared about 140 million years ago, have wings that are supported on the skeleton of the whole arm and hand. But they were furnished with a revolutionary new development –

*Right: with its short, broad wings, the spurwing (*Plectropterus gambiensis*) does not fly well, but it can take off very quickly, an advantage in its wooded habitat. The spur which gives the species its name can be clearly seen.*

feathers. Feathers combine lightness with astonishing strength and elasticity. They can be replaced when they drop out and are also waterproof. Finally, some 50 million years ago, bats made their appearance. The flight membrane of these mammals is stretched between the second and fifth fingers. The first finger remains free.

Right: like the swallowtails, this South American butterfly (Urania leilus) is a good flier, although it belongs to a completely different family. The long wing extensions apparently help to reduce air turbulence.

Below: dragonflies have a highly developed flight apparatus. The front and hind wings can be moved independently of each other. So the insects can hover quite still in the air and also fly off very rapidly.

Although the wing structure appears identical, only the two birds fly in the same way. The brown pelican (Pelecanus occidentalis) on the left, and the rare Galapagos albatross (Diomedea irrorata) on the right, can both fly against the wind. On the other hand, the feather moth Alucita in the centre, with a wingspan of scarcely 2cm, has a quite different method of flying: it more or less floats in the air.

181

THE ART OF FLYING

Below: three phases in the flight of a pelican. The bird at the bottom has brought its feet forwards and is about to land, using its wings as brakes.

Flying is no simple matter. Active flight requires not only wings but also powerful muscles and highly efficient metabolism. Icarus could never have managed it because the muscles of his chest would have been too weak. However, the first animals to take to the air did not engage in active, wing-flapping

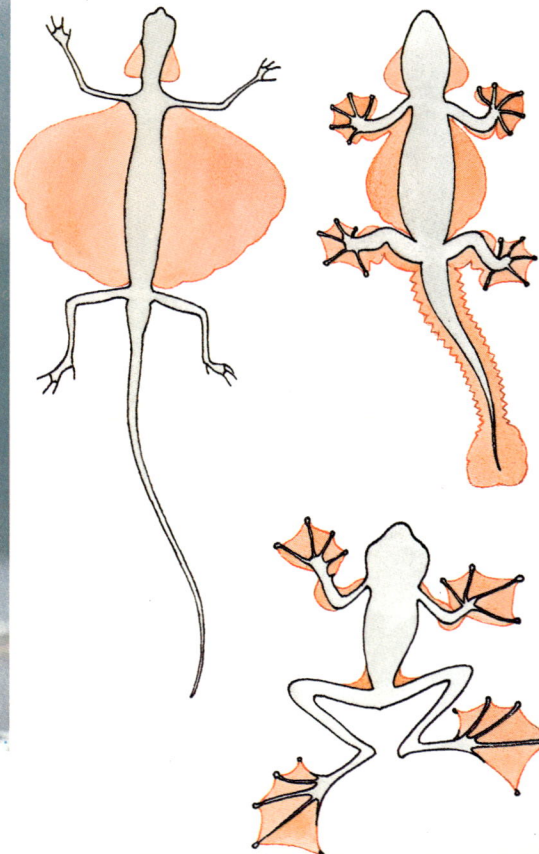

flight, but in gliding. The bird's prehistoric ancestor, *Archaeopteryx*, apparently confined itself to short glides; its breastbone structure suggests the musculature was too weak to support wings capable of true flight.

Apart from birds, there are still certain amphibians and reptiles capable of gliding, sailing across from tree to tree or down to the ground in a more or less straight line. The toe membranes of the rhacophorid frogs and the skin folds of the flying gecko work rather like a parachute, but the flying dragons – small lizards with lateral sails – are even better equipped for gliding. And some mammals have brought the technique to a pitch of perfection. The flying marsupials and squirrels have their flight membrane stretched between the front and hind limbs, and in the flying lemur the tail is also included. With this flight structure they can glide about fifty to a hundred metres. Gliding, however, necessarily involves descending flight. Animals that want to fly straight forward or even upwards must have powerful flapping wings. The frequency of the wing beats depends largely upon the length and weight of the wings. Whereas birds of prey and vultures have a slow majestic pattern of flight, a robin beats its small wings up and down twenty times a second. Hummingbirds flutter at eighty beats per second, thus approaching the rate of insects, about 400 beats per second for certain flies and 1000 for mosquitoes.

FLIGHT PHYSICS

Feathers provide the bearing element of the wings and the aerodynamic form of the body. They smooth out any unevenness and reduce air resistance. These photographs, from below, the front and the side, show the perfect flight form.

Below: a swallow-tailed gull (Creagus furcatus), *showing the large bearing surface. The tail serves to stabilize and steer.*

Birds have no conception of flight physics yet they conform to its laws down to the smallest detail. Their entire body is constructed on aerodynamic principles: the surface of the sternum or breastbone, i.e. the front part, is small, offering as little resistance to the wind as possible; the body is spindle-shaped, with a minimum number of vertebrae, gliding through the air without creating turbulence; and the wings are wonderfully structured for uplift and forward movement.

How has man, in his quest for

Right: a young great black-backed gull, illustrating the small size of the sternum or breastbone, reducing air resistance to a minimum.

mastery of the air, been able to profit from the example of birds? The great Leonardo da Vinci applied himself to the problem almost five centuries ago, discussing the principles of bird flight in his notebooks and making sketches of a flying machine. The theme was taken up again in 1889 when a book was published, entitled *Bird Flight as the Basis of Flying.* The author was Otto Lilienthal, a young engineer who from his

youth had watched storks flying and was obsessed with the idea of doing the same. Seven years after completing his book, he crashed from a height of fifteen metres in an aircraft he had himself constructed. His last words were said to be: 'there must be sacrifices'. Lilienthal did not, in fact, die in vain. From the storks he had managed to discover the vital secret of bird flight: the principle of the curved surface. Bird wings have a convex upper surface, and the wings of all types of aircraft since

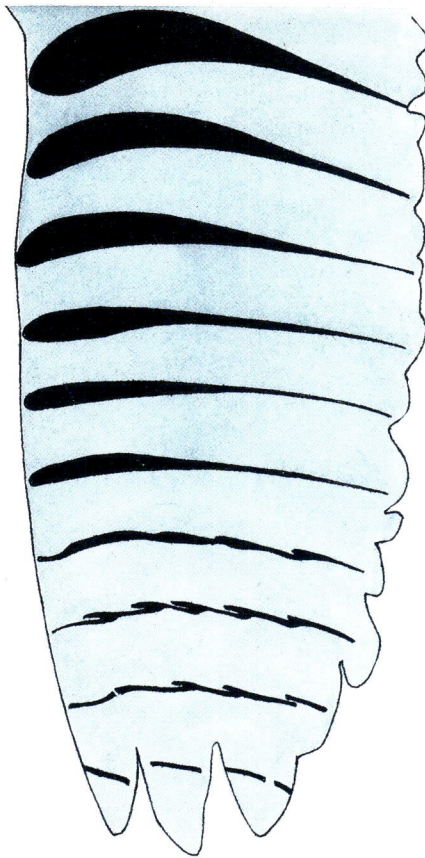

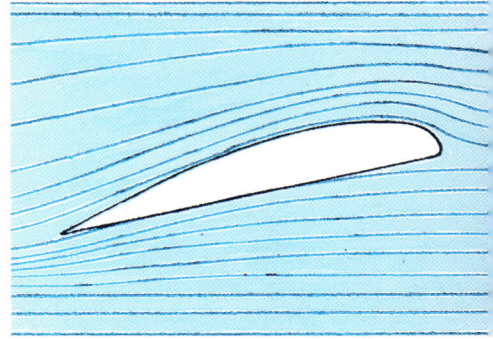

Left and above: profile of a bird's wing, showing its similarity to an aircraft wing. In both instances the air flows faster over the upper surface, creating an upward pressure on the under surface and this lifts the bird or aircraft upwards.

Left: a gannet in flight. The spindle shape guarantees good streamlining.

Lilienthal's day have been built on that model. What happens is that the air parted by the front edge of the wing must travel a longer distance along the upper surface and thus flow faster than on the under surface. The rapid flow on the wing's upper surface produces a reduced pressure which forces the wing upwards, while the air beneath the wing creates an increased pressure and likewise presses it upwards.

The German pioneer Otto von Lilienthal studied the flight behaviour of storks. With varying success, he built several flying machines. In 1896 he crashed in one of these and was killed. His most important discovery was that wings must be curved to produce lift.

185

ANIMAL SENSORS

Right: the scent organ of the female emperor moth is a sac-like structure. Its scent glands produce the sexual attractant which is perceived by the large feathery antennae of the male (facing page) over a distance of several kilometres. On its journey towards the female, the male flies a zigzag upwind course, always seeking the greatest concentration of the scent.

Below: in almost all fishes there is a row of small pits along the flanks, forming the lateral line, which pick up the tiniest vibrations in the surrounding water.

Sense organs and technical sensors basically function in the same way: both receive stimuli from the outside world, convert them into electrical signals and convey them respectively to the brain and to a computer. However, the most modern technological marvels seem primitive and clumsy when they are measured against the extreme sensitivity of biological sense organs.

It is now possible for scientists to investigate the sensory perceptions of animals and to study the responses of their sense organs. One of the most sensitive mechanisms in the whole animal kingdom is the vibration receiver of the house cricket. This organ is situated just below the 'knee' on the insect's

up the oscillations and converts them into a series of electrical nerve impulses. The whole system is enormously sensitive: it can detect minimal oscillations of a billionth of a centimetre; that is about ten times less than the diameter of a hydrogen atom. Man-made vibration detectors are insensitive at much greater orders of magnitude.

The olfactory sense cells on the antennae of a male silk moth are also impressive: under favourable

Above, right: like very many mammals, horses have special olfactory tissue on the gums. When the nostrils are dilated, primarily in preparation for mating, air is sucked in so that it flows over this tissue.

leg and is therefore called the subgenual organ. Because it does not form part of a joint it is highly responsive to vibrations. When the ground vibrates on the approach of a large animal, the whole leg vibrates with it and the vibrations are increased about a hundredfold. The vibration detector picks

conditions a single molecule of the sexual attractant bombycol is sufficient to stimulate this remarkable sense organ.

Almost all adult insects and many larvae possess these and similar organs, not only for smell and hearing, but also for indicating air speed and flight direction.

LUMINOUS ORGANISMS

The deep-sea fauna is still largely unexplored. These animals live in almost complete darkness, at a very constant temperature, which only fluctuates a fraction of a degree over the year. According to the depth the water pressure varies between 50 and 1100 atmospheres. This is why these animals explode when brought to the surface too rapidly.

The drawing (right) shows a variety of abyssal inhabitants. At the top is a ribbon-fish and just below the curve of its tail a luminous squid. Above, right, is a luminescent prawn and at bottom, left, a luminous tube-worm. To the right of it are a dwarf anglerfish, a tower shell and a pagoda shell. At the far right is a Venus flower-basket (a siliceous sponge). In the open water, left to right: a lantern anglerfish, a Grammatostomias fish, a hatchetfish, another deep-sea angler, a dragonfish, a tripod fish and a deep-sea stargazer.

Deep-sea anglers use their light as a lure in the pitch darkness of the ocean abyss.

During the Six Days' War in 1967, an Israeli night patrol on the coast of the Red Sea noticed a mysterious light under water. Assuming that there were enemy frogmen in the vicinity, they threw their hand grenades into the water. To their astonishment, the shore was soon littered with fishes; the bodies were dark, yet the eyes continued to shine brightly and illuminate the surroundings. What they had seen was just a harmless shoal of lanternfishes (*Photoblepharon*).

Bioluminescence, biologically produced light, is a common phenomenon among many marine organisms. Luminous organisms are found both in the deep sea abyss and also in warm surface waters, and they range from bacteria to fishes. Not all the species produce their light themselves: some live in symbiosis with bacteria, which are concentrated in special pits to form light organs. This is the case with the lanternfishes.

The various luminous phenomena thus far investigated in animals are due to a chemical reaction between the polypeptide luciferin and the enzyme luciferase. In this process eighty to

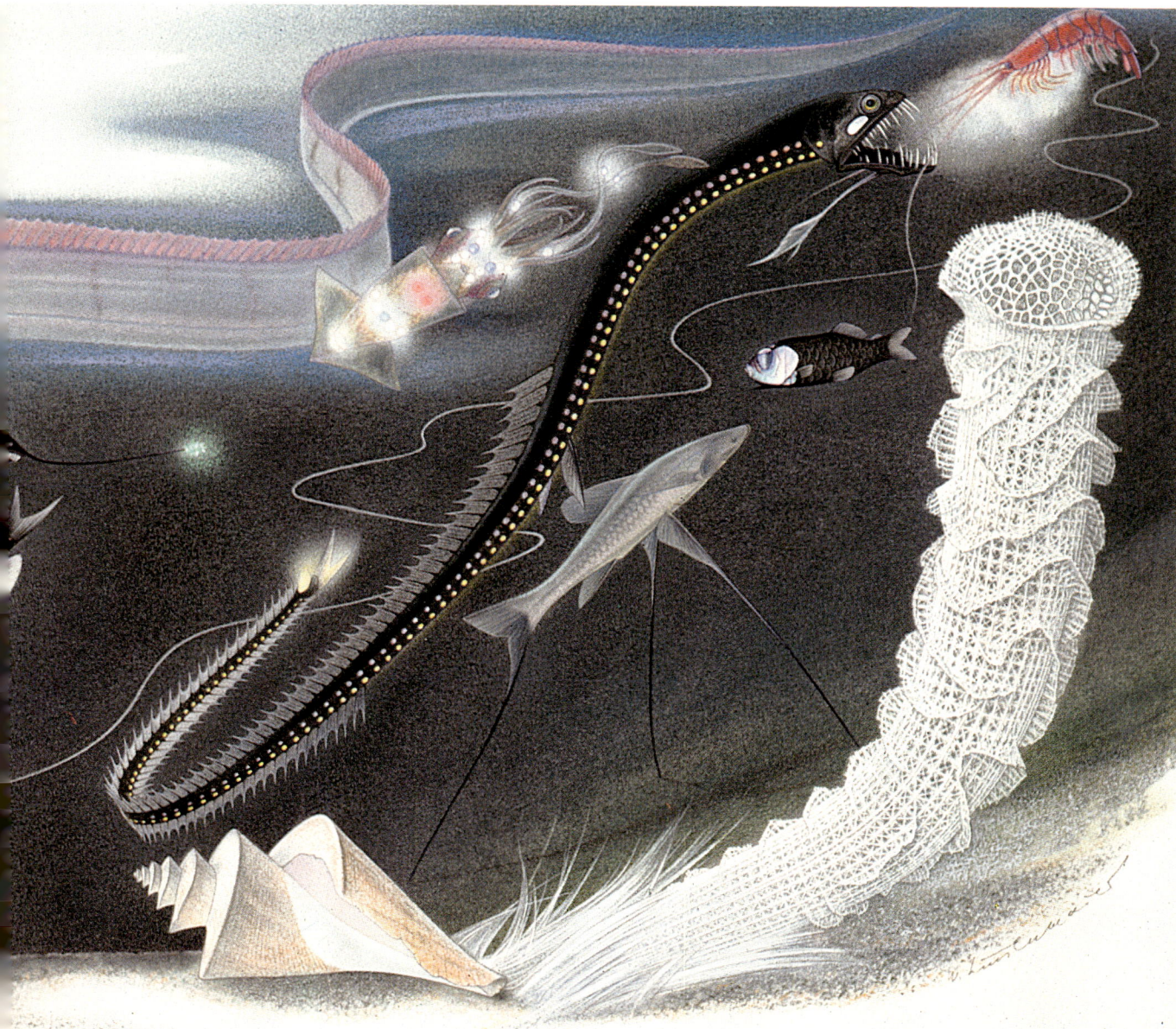

ninety-five per cent of the chemical energy is released in the form of light, the remainder being dispensed as heat. This is an extremely high yield; in an electric light bulb, the ratio is almost the reverse.

The biological importance of bioluminescence varies and is not always clear. In some deep-sea forms it helps to bring the sexes together for the pattern of the light signal differs from one species to another. In other cases, the lights serve to lure prey or emit luminescent clouds to confuse and ward off predators. Fishes and crustaceans of surface waters carry their lights on the underside. In this way they illuminate their dark silhouette which otherwise, seen from below, would be conspicuous against the bright daylight above. In the case of the lanternfish *Photoblepharon*, several of these functions occur in a single species. With the help of special mechanisms it can set its 'searchlight' working. In addition to illuminating the surroundings, enticing prey and communicating with companions, the lights probably serve to frighten enemies when they are switched on suddenly.

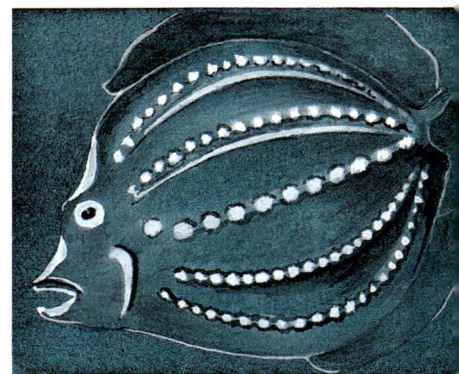

The biological function of light organs is not always understood. In this case the pattern of lights may assist in species recognition.

NATURE AND TECHNOLOGY

Both biologists and technologists are concerned with the fascinating interplay between structure and function. The biologist's starting point is a structure in nature which already has a function, and the objective is to discover its purpose

A particularly interesting area of study is the manner in which animals move, expending as little energy as possible. Dolphins have provided especially valuable information. Scientists have observed that these sea mammals swim

As in a technical docking manoeuvre, the clasping and genital organs of dragonflies are built to fit exactly into one another. This avoids mating between individuals of different species. First, the male lands on the female's thorax and grips her with its tail claspers. The female then bends her abdomen forwards so that her genital opening meets a special apparatus under the male's abdomen. By bending his abdomen forwards, he has already charged this apparatus with sperm. The drawing (right) shows the docking system of an Apollo spaceship.

and how it works. The technologist, on the other hand, knows what needs to be done and must develop the structural mechanism to fulfil the desired function. Collaboration between these two forms of specialized science can be very fruitful, and in recent years the development of a new scientific discipline known as bionics has brought biologists and technologists together. One scientist in this field has broadly defined the aim of bionics as learning the secrets of nature and from them establishing the principles on which to build technical structures.

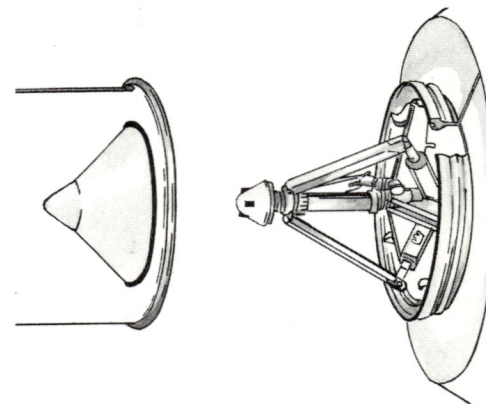

much faster than would, in theory, appear likely on the basis of muscle power alone. Hitherto unknown principles must therefore be involved. The first secret to be discovered was that the skin of dol-

phins is so structured as to be capable of reducing turbulence along the whole length of the body thus lessening resistance to the water. The second was that the dolphin's torpedo-shaped body is ideally constructed, hydrodynamically, for rapid underwater swimming. One bionic engineer has in fact recommended that this shape should be used for aircraft fuselages, making possible larger, roomier planes without the problem of increased air resistance. The third secret of dolphin movement is that it combines that of the fishes and of other sea mammals. They propel themselves forward by moving the tail from side to side, like fishes, and up and down, like sea mammals, so compensating for the resistance inevitably encountered by a body moving through water. In this respect, their propulsion method is more effective than a ship's propeller.

Dolphins have an extraordinarily well streamlined body. This shape might well be emulated in the construction of aircraft fuselages.

Hummingbirds beat their wings up to 80 times per second. Analysis by high-speed photography shows that the wing tips describe a horizontal figure of eight, a flight method very close to the rotor of a helicopter.

SKY NAVIGATION

Before an animal can use the celestial bodies for orientation, it has to know their movements exactly.

At night birds navigate by the stars, but to do this they must first learn their positions and patterns. Experienced birds only need a small area of open sky for direction-finding.

When we go out to buy a loaf of bread from the local baker, we do not need to orientate ourselves by the sun or stars. We are familiar with the neighbourhood and therefore know exactly where to cross and to turn in order to reach our destination. When we get near, our noses may give us further information. Animals also know their immediate surroundings and find their direction according to certain recognized landmarks and scents. Indeed, domestic cats and dogs are much better direction-finders than we are ourselves.

There are, however, situations in which such knowledge is useless. If we are abroad, we may need to consult a guidebook or ask for directions. Many animals wander through areas they do not know. Others live in very uniform surroundings with scarcely any landmarks. Consider, for example, birds flying over the ocean or ants travelling across a flat desert. In such cases, apart from the earth's magnetic field, which almost certainly provides clues, points of reference may be supplied by the sun, moon and stars or by the polarization patterns of the sky, due to filtration of sunlight. Investigations have shown that such aids to orientation are definitely used by certain animals. Many of them can navigate by using the sun as a compass. Migrating birds, which often have to fly enormous distances, and at night, use the stars to find their way. They even do so when the sky is partly or wholly overcast, since they know by heart the positions of the various constellations. Insects cannot recognize the stars but they can, at least by day and with an almost completely overcast sky, perceive the polarization pattern, invisible to us, of sunlight in the atmosphere, and from it reckon the actual position of the sun.

It is no exaggeration, therefore, to say that animals could be compared to astronomers, especially bearing in mind that the sun, moon and stars are not stationary. To make use of these bodies rather in the manner of a compass, animals would have to know exactly their daily and annual movements across the sky. This is precisely what many animals have managed to do; and this explains, at least in part, how migrating birds can find their way so unerringly to their winter quarters and back to their breeding sites.

Left: many animals use the sunshine for orientation. These include not only migrating birds, but also mammals, reptiles, amphibians, insects and even crustaceans. To achieve this the animals know exactly where the sun is at a certain time of day. They must also have a means of measuring time: a so-called internal clock. The picture shows an arctic tern which flies on its annual migration from the high arctic to the antarctic zone and back, covering some 30,000 kilometres.

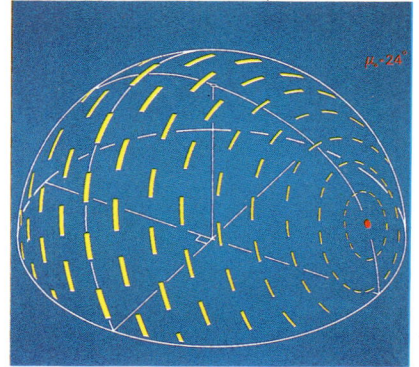

Above: in the atmosphere, incident sunlight falling obliquely is polarized. This produces a pattern, invisible to us, which moves across the sky. It is used for orientation particularly by insects and crustaceans.

For many years a research team from the University of Zurich has been studying the orientation behaviour of the desert ant (Cataglyphis bicolor). The research area in the Tunisian Sahara has scarcely any landmarks. So the ants have to orientate solely by the sky.

Colour-marked research ants were able to find their way even if the sun was artificially obscured. The conclusion is that they determine direction by the sky's polarization patterns.

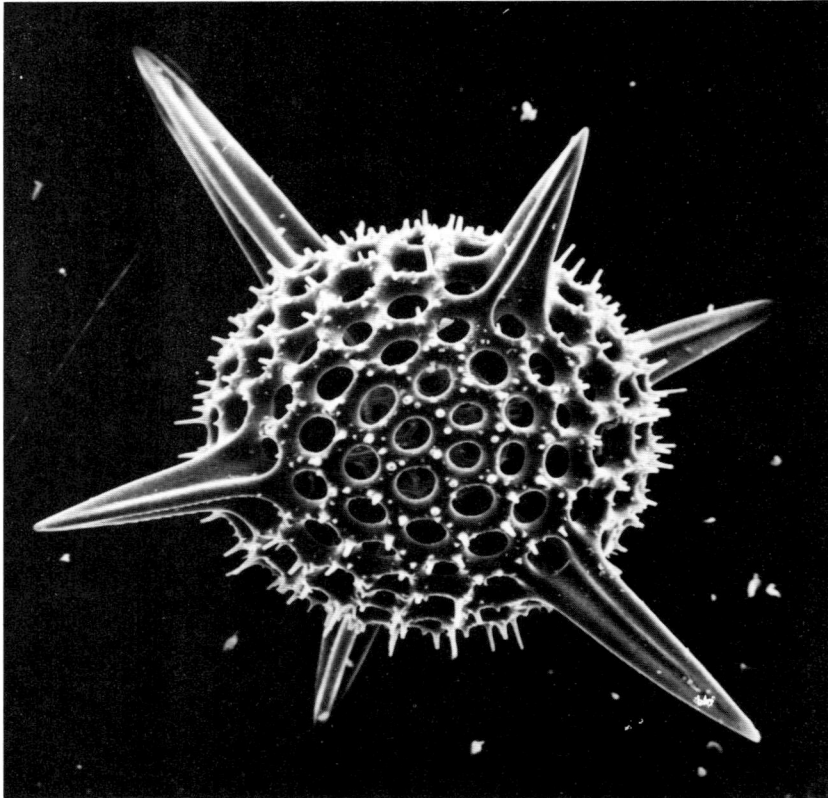

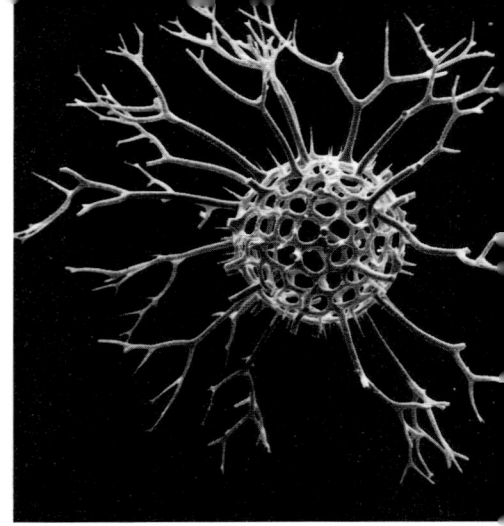

Above and top right: radio-larians are tiny free-floating marine animals. The spiky processes and lightweight oil globules enclosing them enable the creatures to remain, without actively swimming, at the desired depth. These fossil radiolarian skeletons come from a deep-sea bore by the research ship Glomar Challenger.

Right: the calcareous shell of Nautilus *consists of a series of chambers whose content of gas and liquid can be regulated through the connecting tubular siphunculus. The animal sinks or rises according to the gas content of the chambers.*

Lightness and solidity are essential elements in aircraft construction; and the same is true of the animal kingdom, because every extra unnecessary gram costs energy, and this can only be replaced from the restricted supply of food. To be lightly built is particularly important for animals that fly and also for those that float in the surface water of the sea. The less they weigh, the greater their stamina and powers of manoeuvre.

In the oceans, the uppermost water layer is illuminated and permits the growth of innumerable small plant organisms. This phytoplankton provides the basic food for the animal plankton or zooplankton to which the foramini-ferans and radiolarians belong. These are microscopically small animals with a very complicated, often perforated external skeleton, calcareous in the foraminiferans, usually siliceous in the radiolarians. The long skeletal processes and the oil globules that enclose them provide buoyancy. To sink into the lightless depths would be fatal.

The molluscs of the species *Nautilus* have a shell consisting of numerous chambers connected by a tube that secretes gases. According to the quantity of gas in the chambers, the animal can control its ascent and descent. The principle is similar to the bony fish's swim bladder – a thin sac filled with gases. The amount of air in the bladder helps the fish adjust to changing water pressure and controls its depth, as a submarine does by adjusting its ballast.

The largest flying bird cannot be heavier than about 15kg. Beyond this limit, flight would

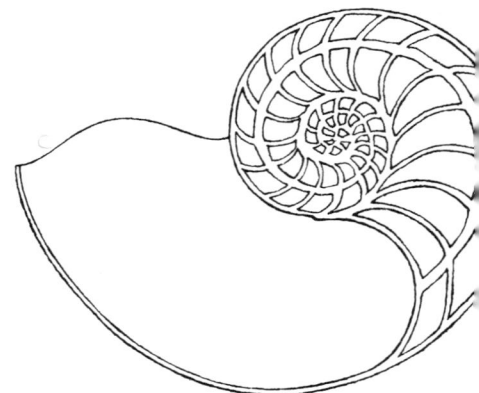

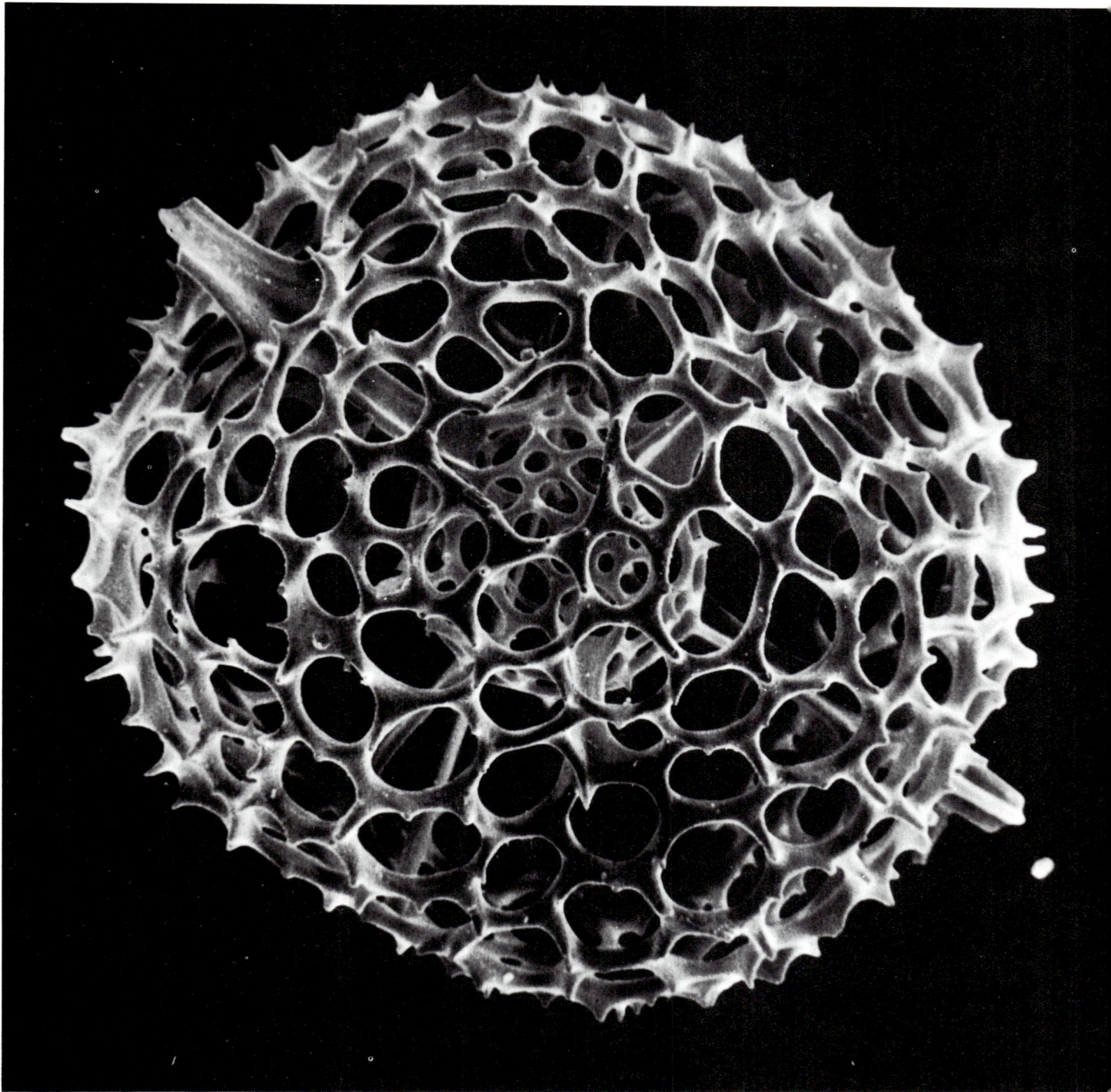

entail too much effort. To reduce their weight, birds have a system of air sacs, which extend into the largely hollow bones. A bird's body is subject to heavy loads, and must also be very stable. So the humerus bone, for instance, has a system of buttress-like processes that is comparable to the struts of an aircraft wing.

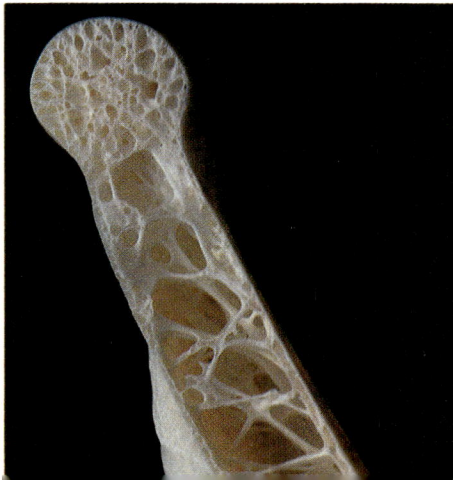

Above: some radiolarian skeletons consist of two spheres, one inside the other.

Left: longitudinal section through the humerus of a pelican. The struts combine solidity with extreme lightness, always necessary for flight.

HUNTING
WITH
ULTRASOUND

Below: in contrast to the ves-pertilionid bats, the horseshoe bats emit sounds through the nostrils. These ultrasounds are guided in the direction of flight by a complicated nasal process. The illustration is of the greater horseshoe bat (Rhinolophus ferrumequinum).

Above right: flying vespertilionids like this brown long-eared bat (Plecotus auritus) emit through the open mouth about 10 ultrasonic orientation sounds per second, each lasting about 100 milliseconds. When prey or an obstacle is encountered, the series of sounds increases to 100 per second. At the same time the individual pulse shortens so that their echo is always heard in the intervals between the emissions.

At the time of the French Revolution, the Italian abbot Lazzaro Spallanzani and the Swiss naturalist Ludwig Jurine conducted experiments with bats. Their investigations showed that bats were capable of flying, and even of avoiding obstacles, without using their eyes. On the other hand, bats with their ears blocked completely lost the ability to orientate. However, when Spallanzani inserted small tubes into the wax block, thus keeping the entrance to the ear open, the bats regained their ability to find their way around in complete darkness.

Scientists of the present century have furnished the answer to this puzzle by demonstrating that bats possess a highly developed system of acoustic radar. They emit closely packed ultrasounds which, at a distance of 5–10cm from the tip of their nose, give out a volume of sound comparable to that of a pneumatic drill. We are incapable of hearing this din because our ears

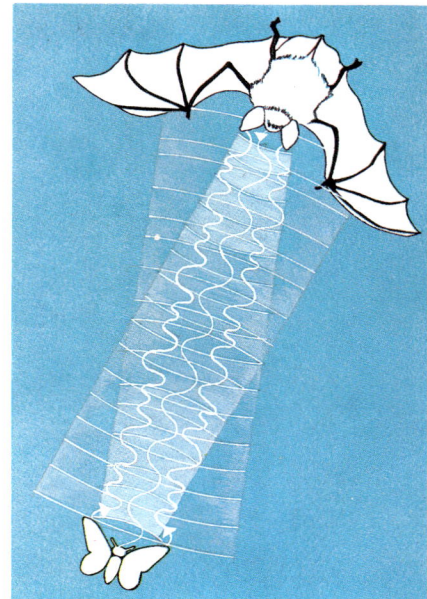

Above: rough or smooth, small or large surfaces modify the ultrasounds in different ways and thus lead to different echoes. The strength of the returning signal gives information on the distance of prey, and also on its speed and flight path.

can pick up frequencies of only about 20 kilohertz, whereas the sound waves emitted by bats have frequencies ranging from 20 to 160 kilohertz. They can hear their own ultrasounds very well and find their whereabouts from the weak echoes thrown back off solid objects by their own cries.

At first sight it might appear that human radar is superior to the animal system: after all, radar waves reach the speed of light, whereas bat signals only propagate with the velocity of sound. However, every radar expert knows the difficulty of detecting an aircraft that is flying close to the ground: its echo blends with those from the earth's surface. Bats are able to master this problem brilliantly: they can even detect the smallest insects sitting on leaves. Fish-eating bats have no trouble in perceiving ripples on the water surface caused by swimming fishes, and can even distinguish between different fish species.

FLASHING
SIGNALS

The light signals of fireflies serve to bring the sexes together. In regions where several species exist together, it is essential that such signals should be different and specific to a particular species. The picture on the right, based on the work of James Lloyd, an American researcher, illustrates the flash sequences of the different species of firefly. Normally the male flies over an area and flashes his own signal. The female, hidden in the grass, then gives the appropriate answering signal, whereupon he flies down and mates. There is, however, one genus of predatory firefly (Photuris) in which the female imitates the reply signal of another species (Photinus) and lures the unsuspecting male to the ground where she consumes him. This is shown in the smaller picture above.

In South-east Asia certain beetles, commonly known as fireflies, sit in thousands on the trees, simultaneously flashing their lights on and off. The effect is that of a giant Christmas tree.

Most species of firefly send out their signals singly. Usually the males fly over a meadow and advertise their presence to the females by characteristic flashes which vary from species to species. Even when different species occur alongside one another, a female waiting in the grass knows exactly when the 'right one' approaches. Only then does she flash back. For the male this is the signal to land and start mating. The signals of either sex are given out at exactly regular intervals.

Certain predatory female fireflies of the genus *Photuris* feed on smaller fireflies of the genus *Photinus*, imitating the enticement signals of the females of their prey. The unsuspecting *Photinus* male thinks he has found a female of his own species ready to mate. Instead he flies directly to his death.

There are, in fact, numerous species, both in the Old and New World; and their light organs, like those of luminescent deep-sea fishes, are produced by the chemical reaction of the enzyme luciferase on the polypeptide luciferin. In South America certain peasants of Ecuador have become skilful at catching the glowing beetles; they put them in small receptacles which they then sell to be used as nightlights and reading lights.

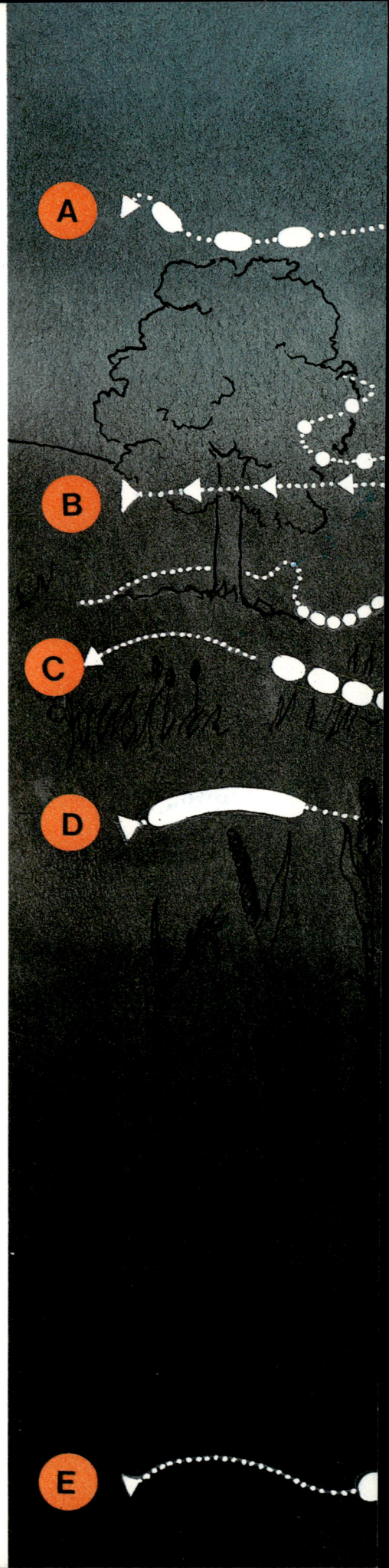

BIBLIOGRAPHY

Attenborough D. *Life on Earth*: Collins, 1979
Baker R.R. *Evolutionary Ecology of Animal Migration*: Hodder & Stoughton, 1978
Barrett J. *Life on the Seashore*: Collins, 1974
Bristowe W.S. *The World of Spiders*: Collins New Naturalist, London, 1947
Burton J. *The Book of the Year: A Natural History of Britain through the seasons*, Warne, London, 1983
Burton J. & Taylor K. *Nightwatch*: Michael Joseph, London, 1983
Burton M. *The Sixth Sense of Animals*: Dent, London, 1973
Burton M. & R.W. *Inside the Animal World: an encyclopedia of animal behaviour*, Macmillan, London, 1977
Burton R.W. *Bird Behaviour*: Granada, 1985
Colinvaux P. *Why Big Fierce Animals are Rare*: George Allen & Unwin, 1978
Cott H.B. *Looking at Animals*: Collins, 1975
Eltringham S.K. *Life in Mud and Sand*: English Universities Press, 1971
Green J. *A Biology of Crustacea*: Witherby, London, 1961
Griffin D.R. *Listening in the Dark*: The Acoustic Orientation of Bats and Men, Yale University Press, New Haven, 1958
Hardy A. *The Open Sea*: Collins New Naturalist, 1956
Herring P. & Clarke M.R. *Deep Oceans*: Arthur Barker, 1971
Imms A.D. *Insect Natural History*: Collins New Naturalist, London, 1947

Lane F.W. *Nature Parade*: Jarrolds, London, 1939
Lane F.W. *Animal Wonderland*: Country Life, 1948
Lorenz K. *King Solomon's Ring*: Methuen, 1952
Louw G.N. & Seely M.K. *Ecology of Desert Organisms*: London, 1982
Lythgoe J.N. *The Ecology of Vision*: Clarendon Press, Oxford, 1979
McFarland D. (ed) *The Oxford Companion to Animal Behaviour*: Oxford University Press, 1981
Marshall N.B. *Developments in Deep Sea Biology*: Blandford, 1979
Mead C. *Bird Migration*: Country Life, 1983
Norman J.R. & Greenwood P.H. *A History of Fishes*: Benn, London, 1963
Perry R. *Mountain Wildlife*: Croom Helm, 1981
Schmidt-Nielsen K. *How Animals Work*: Cambridge University Press, 1972
Stonehouse B. *Animals of the Arctic: the Ecology of the Far North*, Eurobook, 1971
Stonehouse B. *Animals of the Antarctic: the Ecology of the Far South*, Peter Lowe, London, 1972
Tinbergen N. *The Herring Gull's World*: Collins New Naturalist, 1953
Tinbergen N. *Curious Naturalists*: Country Life, 1958
van Lawick-Goodall J. *My Friends the Wild Chimpanzees*: National Geographic Society, 1967
Wilson E.O. *The Insect Societies*: Belknap Press of Harvard University Press, Cambridge, 1971
Wilson E.O. *Sociobiology – The New Synthesis*: Belknap Press of Harvard University Press, Cambridge, 1975

INDEX

PICTURE CREDITS

after *Alcock*, bottom); 30 (centre right *after Mühlenberg and Maschwitz*); 58 (bottom left *after Anzenberger*); 59 (bottom left *after Wilson*); 76 (left *after Kirschfeld*); 95 (top and bottom *after Alcock*); 112 (top right *after Hölldobler*); 114 (bottom *after Henry*); 116 (at very top *after Eibl-Eibestedt*); 117 (top *after Eibl-Eibestedt*); 120 (top right *after Wilson and Lüscher*); 120/121 (bottom *after Kotschnig*); 121 (top *after Skaife and Landry*); 122 (left top and bottom *after von Lawick-Goodall*); 123 (centre right and bottom *after von Frisch*); 124/125 (top *after Partridge*); 125 (bottom after *Pilleri and Knuckey*); 126 (top *after De Vore and Berill*); 147 (centre *after Baker*); 168 (bottom centre and right); 174 (top left *after Wheeler*); 175 (bottom); 180 (left *after Wellnhofer*); 182 (bottom right); 185 (top left and right *after Rüppell*); 188 (bottom); 189 (bottom); 190 (bottom right *after Nachtigall*); 194 (bottom right); 197 (right *after Habersetzer*); 199 (*after Lloyd*).
IKAN Frankfurt/J. Debelius: 34/35
Isenhart, H.-H. and *Bührer*, E.M., *Das Königreich des Pferdes*, 1969, C.J. Bucher, Lucerne and Frankfurt-am-Main:
152 (top left).
JACANA, Paris:
Y. Arthus-Bertrand: 100 (bottom);
Coll. Varin-Visage: 118 (right); 138 (top centre);
F. Danrigal: 3; 62 (top right);
Ermie: 61 (left); 172 (centre);
J.P. Ferrero: 100 (2nd from bottom); 104; 153 (bottom, second from right);
Frédéric: 150 (left);
F. Gohier: 44 (bottom); 46/47; 148/149;
J.M. Labat: 105; 173 (right);
Mero: 61 (right);
C. Pissavini: 163 (bottom left);
Suinot: 107 (centre); 161 (left);
J.P. Varin: 27 (centre); 31; 32; 33 (bottom); 153 (bottom right); 174 (top right);
A. Visage: 103 (top left); 153 (bottom 2nd from left);
F. Winner: 62 (left and centre right);
G. Ziesler: 101 (2nd from bottom).
Killmeyer, Franz, Tullnerbach: 16.
Klages, Jurg, Zurich: 103 (centre right).
Knapp, Egon, Nuehausen:
9 (bottom right); 12 (centre); 14 (left); 23; 26 (bottom); 36 (3rd row on left from top, and 3rd row on right from bottom); 37 (top left); 50; 53 (top); 57 (two at bottom); 63 (left hand row, 2nd and 3rd from top); 77; 79 (left top and bottom); 82 (bottom left); 83 (bottom right); 85; 94 (top); 98 (top and centre right); 99 (top); 108 (bottom centre and right); 109 (bottom); 119 (left); 128 (top centre and right, and bottom); 135 (3rd from top); 168 (top left and bottom); 174 (two at

bottom); 168 (top right); 190 (top).
Krebs, Albert, Winterthur:
36 (bottom left); 83 (bottom centre); 94 (bottom); 98 (left and bottom); 99 (bottom); 128 (top left).
Leutert, Dr. Alfred, Schaffhausen:
12 (right); 102 (top).
Linsenmaier, Walter, Ebikon:
2; 24; 25 (centre left and bottom); 37 (bottom); 51; 64; 65; 80 (top, bottom left and centre); 81 (left); 116 (bottom); 118 (left); 123 (top and centre left); 129, 131, 138 (top); 188/189 (top).
Lloyd, Prof. James E., Gainesville;
198 (left).
Meier, Max, Zurich:
18 (top); 19; 72; 73; 74; 75; 101 (bottom).
Muller, Marten, Affolfern a. A.:
172 (right).
Naturgeschichte des Tierreichs, after Prof. G.H. von Schuberts, Esslingen near Stuttgart, 1886:
142 (top right); 143 (top, the two at left, and centre far right); 144 (bottom left).
Museum of Natural History, Berlin:
182 (top right).
Official US Navy Photograph:
192 (bottom).
Photo Researchers, New York:
B. Brake: 139;
F. Erize: 57 (top).
Photoswissair, Zurich:
191 (centre right).
PKG, Munich:
Menz: 101 (top);
Pfletschinger: 68.
Popperfoto, London: 56 (right).
Rasa, Dr. Anne E., Bayreuth: 127 (top left).
Rüppell, Prof. Dr. Georg, Braunschweig: 141 (top).
Schinz, H.R., *Naturgeschichte und Abbildungen der Säugetiere, Zurich*, 1824:
145 (top).
Schönauer, Peter, Emmenbrücke:
191 (bottom right).
Silvestris Fotoservice, Kastl/Jacana: 160.
STACK, TOM & ASSOCIATES, Colorado Springs/Warren Garst: 21 (top).
Steel, R. and *Harvey*-A.P., *The Encyclopedia of Prehistoric Life*, 1979: McGraw-Hill Book Co., New York: 152 (top right).
Steiger, M: 192 (top).
STERN Hamburg/Jürgen Gebhardt: 141 (bottom).
Stihl, Hj. Buchs: 114 (centre); 159 (bottom).
Wehner, Prof. R., Zurich: 193 (top right and bottom right).
Weidemeider, Patrick, Zurich:
9 (top); 10 (bottom right, at top); 134 (bottom); 135 (bottom); 172 (left); 183 (bottom).
Zellweger, Elsbeth, Zilhlschlacht:
42; 43 (bottom).